THE **CHASE**...

By the time Rosetta caught up with comet Churyumov-Gerasimenko, the craft had completed almost five circuits of the inner Solar System and covered a distance of more than 6,500 million km (4,038 million miles). Along the way, the craft used the gravitational pull of Earth and Mars to accelerate from its launch speed of 26,000 kph (16,155 mph) to the 135,000 kph (83,885 mph) it needed to chase down the comet.

14 **Towards the Sun**
In December 2014, *Rosetta* accompanies the comet as it travels towards the Sun. As the comet warms, its ices "sublimate" (pass straight from solid to gas) and are ejected at supersonic speeds. *Rosetta* records and studies these changes.

1 **Blast off!**
Rosetta launches from Kourou, French Guiana, onboard an Ariane 5 rocket in March 2004.

2 **Earth slingshot 1**
A year after launch, the craft uses Earth's gravity to accelerate.

3 **Mars slingshot 1**
February 2007.

Sun

Earth

1st orbit

4th orbit

5th orbit

Asteroid Lutetia

Comet's orbit

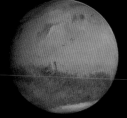

4 **Earth slingshot 2**
November 2007.

5 **Asteroid Steins**
The craft passes within 800 km (497 miles) of the 5 km (3 mile) wide asteroid, and collects information and images in September 2008.

7 **Asteroid Lutetia**
In July 2010, *Rosetta* passes the 100 km (62 mile) wide asteroid Lutetia at a distance of about 5,000 km (3,106 miles).

6 **Earth slingshot 3**
The final Earth slingshot happens in November 2009.

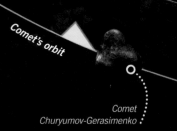

Comet Churyumov-Gerasimenko

SCIENCE BUT NOT AS WE KNOW IT

CUTTING-EDGE CONCEPTS MADE SIMPLE

WRITTEN BY

BEN GILLILAND

CONSULTANT

JACK CHALLONER

DK LONDON

Senior Project Editor Steven Carton
Senior Art Editor Stefan Podhorodecki
Editor Francesca Baines
Editorial Assistant Charlie Galbraith
Designers Sheila Collins, Mik Gates
Managing Editor Linda Esposito
Managing Art Editor Michael Duffy
Jacket Editor Maud Whatley
Jacket Designer Mark Cavanagh
Jacket Design Development Manager Sophia MTT
Producer, Pre-Production Luca Frassinetti
Producer Gemma Sharpe
Publisher Andrew Macintyre
Publishing Director Jonathan Metcalf
Associate Publishing Director Liz Wheeler
Design Director Phil Ormerod

DK INDIA

Editor Priyanka Kharbanda
Art Editors Supriya Mahajan, Heena Sharma
Assistant Editor Deeksha Saikia
Assistant Art Editor Tanvi Sahu
DTP Designers Vishal Bhatia, Nityanand Kumar
Picture Researcher Deepak Negi
Senior DTP Designer Harish Aggarwal
Jacket Designer Vikas Chauhan
Managing Jackets Editor Saloni Talwar
Pre-production Manager Balwant Singh
Production Manager Pankaj Sharma
Managing Editor Kingshuk Ghoshal
Managing Art Editor Govind Mittal

First published in Great Britain in 2015
by Dorling Kindersley Limited
80 Strand, London WC2R ORL

Copyright © 2015 Dorling Kindersley Limited

A Penguin Random House Company
2 4 6 8 10 9 7 5 3 1
001–275156–04/15

ISBN 978-0-2411-8419-6

Printed in China

Discover more at
www.dk.com

CONTENTS

MYSTERIOUS UNIVERSE

HOW BIG IS THE UNIVERSE?

THE ★ STAR THAT REDREW THE COSMOS

⋆ EXPANDING ⋆ UNIVERSE ⋆

WELCOME TO THE MULTIVERSE

CATCH UP WITH STELLAR SPEED DEMONS

WE'RE ALL DOOMED! ☠

MEET THE SMELLY DWARF

THE SEARCH FOR ALIEN LIFE

? MERCURY'S SECRETS

MYSTERIOUS UNIVERSE

HOW TO CATCH A COMET

THE HOSTILE BLUE PLANET

SATURN'S AMAZING RINGS

THE SPACE ROCK THAT "KILLED" PLUTO

HOW BIG IS
THE UNIVERSE?

THAT BIG WHITE SPLODGE ON THE RIGHT IS THE SPIRAL GALAXY NGC 1345
(we will call this one Terry). Terry lives quite close to our very own galaxy, the Milky Way – you might say that they are neighbours. However, closeness is a very relative term indeed. Compared to the **overwhelming vastness** of the Universe, Terry lives just a few doors down the road.

But where does he live in relation to you and me? After all, if you live in a block of flats in New York City, a few doors down is just a few paces along the hallway. However, if you live in the middle of the Mojave Desert, reaching your nearest neighbours could mean having to hop onto your scooter for a trip of several kilometres. Given that the scale of the Universe is more like that of the Mojave than that of the Big Apple, you can be sure **Terry does not live as close as the image implies.**

Let us apply some **sense of scale** to the image. The bright star in the image (the one with the word "star" pointing at it) does not actually live in Terry's house – it actually lives in our house (the Milky Way) – so it must be pretty close. But our house is pretty big, so the star is not as close as you might guess. In fact, that pinpoint of light is probably a few thousand light-years away, and that is still a long way off indeed – because **a light-year** is the distance that a photon of light, shooting along at an impressive **18 million kilometres a minute (11 million miles a minute)**, can travel in a year.

Even if it is as close as a thousand light-years away, that star is still at least

9,500 million million kilometres (5,900 million million miles) away – that journey would take you about **10 billion years** to complete on your scooter (provided you travel 24 hours a day at the heady speed of 100 kph (62 mph)).

Peer a little deeper into the image and you can see lots of small galaxies that seem to be crowding around Terry. Of course, these galaxies only appear much smaller because they live much further down the road than Terry – perhaps hundreds of millions of light-years further down the road.

It is hard (perhaps impossible) for the human brain to comprehend distances of this magnitude, but (in cosmic terms) **we have still barely left the end of the road.** To peer beyond the road and out of town

This is how distant galaxies crowding Terry look when enlarged

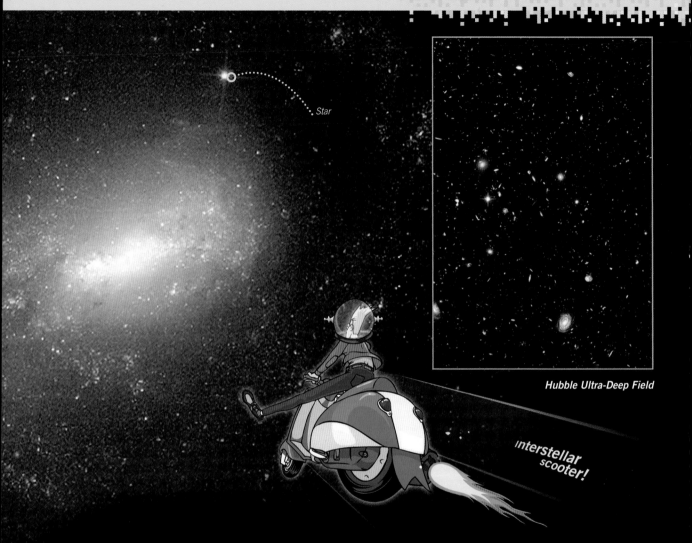

Star

Hubble Ultra-Deep Field

interstellar scooter!

you need a different image. The portrait of Terry was taken by the Hubble Space Telescope using an exposure of about half an hour and, just like using a normal camera, the longer you expose the "film" to light, the more light you gather, and the more light you gather, the fainter the objects you can see.

The image in the top right corner is the **Hubble Ultra-Deep Field.** It is perhaps one of the most profound images ever captured. The image is the result of an exposure amounting to 1 million seconds

(11-and-a-half days). Now, when you consider that Terry and his distant neighbours were revealed after a 30-minute exposure, imagine what is revealed after an exposure of more than 11 days. There are 10,000 galaxies visible in this image and the **most distant is located 13 billion light-years away** – that is a journey of 140 million billion years on your interstellar scooter (you might want to pack a sandwich). However, even though you have to travel well beyond the end of the road, out of town and far out into the distance, even these

galaxies really only sit on the cosmic horizon – the Universe extends far deeper still.

The Universe is not infinite – it does have its limits – but because it is expanding, you could never hope to travel to its end. Even if you were to soup up your scooter to be able to travel at the speed of light, you would still be left playing eternal catch-up with the Universe's ever-expanding frontiers. **Suddenly, Terry doesn't seem so far away!**

HOW BIG IS BIG?

When you see an astronomical object afloat in the blackness of space, without a familiar object nearby to provide some scale, it's difficult to appreciate just how big big can be. So we'll start with Earth – home to some 7 billion humans... so quite big – and go from there...

EARTH
Diameter: 12,756 km
(7,926 miles)

Earth

JUPITER
Diameter: 142,984 km
(88,846 miles)

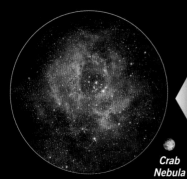

Crab Nebula

ROSETTE NEBULA
Diameter: 1,230 trillion km
(764.2 trillion miles)

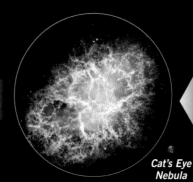

Cat's Eye Nebula

CRAB NEBULA
Diameter: 104 trillion km
(64.6 trillion miles)

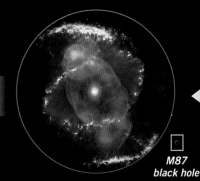

M87 black hole

CAT'S EYE NEBULA
Diameter: 3.78 trillion km
(2.3 trillion miles)

Rosette Nebula

SMALL MAGELLANIC CLOUD (GALAXY)
Diameter: 66,200 trillion km
(41,134 trillion miles)

Small Magellanic Cloud

MILKY WAY (GALAXY)
Diameter: 1.14 million trillion km
(708,363 trillion miles)

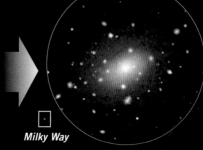

Milky Way

IC 1101 (GALAXY)
Diameter: 53 million trillion km
(32.9 million trillion miles)

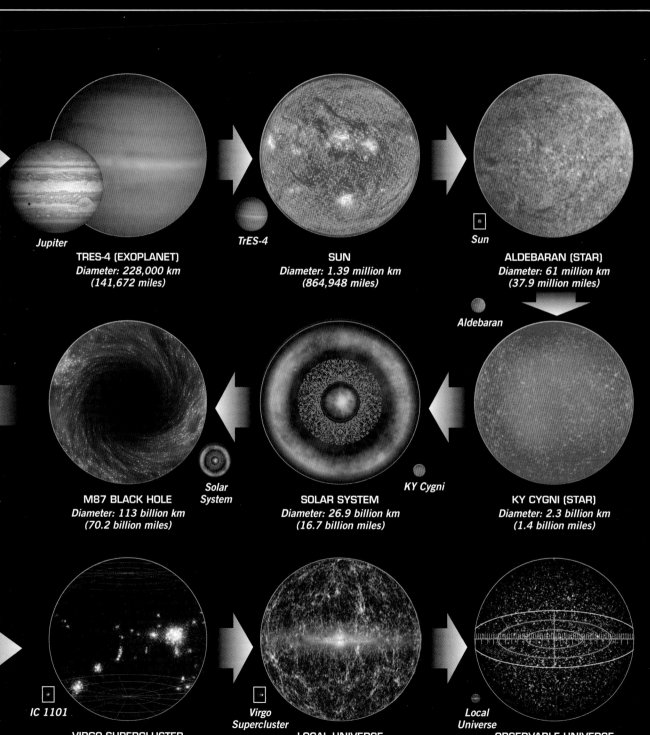

Jupiter

TrES-4

TRES-4 (EXOPLANET)
Diameter: 228,000 km
(141,672 miles)

Sun

SUN
Diameter: 1.39 million km
(864,948 miles)

Sun

ALDEBARAN (STAR)
Diameter: 61 million km
(37.9 million miles)

Aldebaran

*Solar
System*

KY Cygni

M87 BLACK HOLE
Diameter: 113 billion km
(70.2 billion miles)

SOLAR SYSTEM
Diameter: 26.9 billion km
(16.7 billion miles)

KY CYGNI (STAR)
Diameter: 2.3 billion km
(1.4 billion miles)

IC 1101

*Virgo
Supercluster*

*Local
Universe*

VIRGO SUPERCLUSTER
Diameter: 1.04 billion trillion km
(646 million trillion miles)

LOCAL UNIVERSE
Diameter: 24.6 billion trillion km
(15.28 billion trillion miles)

OBSERVABLE UNIVERSE
Diameter: 880 billion trillion km
(546.8 billion trillion miles)

THE STAR THAT REDREW
THE COSMOS

AT THE START OF THE 20TH CENTURY, astronomers thought they knew what the Universe was all about. It was an island of light, afloat alone in the dark, infinite sea of existence. Measuring about **100,000 light-years** across, it contained about 100 million stars – a fixed, unchanging, and eternal raft of stars sometimes called the Milky Way.

Then, on 4 October 1923, from a dark mountainside in California, USA, a discovery was made that would **redraw the map of the cosmos.** It led astronomy down a path that no one could have imagined or predicted, and would eventually lead the way back **13.8 billion years** to the birth of the Universe itself.

From the time of the ancient Greeks to the Age of Enlightenment (nearly 2,000 years later), mainstream belief had it that the **full extent of the Universe was contained within a series of celestial spheres,** which encased the focal-point, and jewel, of creation – planet Earth. In the 16th century, Polish astronomer Nicolaus Copernicus convinced many people that Earth was not at the centre, but was in orbit around the Sun – and **discoveries made after the introduction of the telescope to astronomy in 1609 convinced many of the doubters.** The centuries-old map of the heavens was torn up, and the Sun found itself suddenly promoted to the position of "centre of the Universe". But, within just a few short years, its promotion lost its significance and **the Sun was relegated to being just one star of many tens of thousands** in a new-model Universe called the "Milky Way".

Once astronomers had the Universe's "true" size and nature sketched out, the only thing astronomers really had left to do was find stuff floating around within its confines. **They sought new stars, asteroids, or comets,** to which, like the great explorers of old, they could attach their names for the advancement of science and not at all for fame or other such vainglorious pursuits (well, maybe for a bit of glory).

In the latter half of the 18th century, one of the cosmological explorers was a French chap called Charles Messier. Messier was obsessed with finding comets (he would discover 13 in all, with six more co-discoveries) but, as he scanned the heavens, he kept stumbling across **strange fuzzy objects** – fluffy, cloud-like blobs that looked a bit like comets but which did not

seem to move. To avoid confusing them in the future, Messier compiled a catalogue of these **"nebulae"** and, by the time he died in 1817, he had charted the locations of 103 of them.

But what were they? Why did some appear to be shapeless apparitions, while others formed spirals? Were they, as many believed, just insignificant clouds of gas and random groups of stars that floated around inside the Milky Way? Or were they, as the great German-British astronomer William Herschel suggested, **unique island Universes located beyond the limits of the Milky Way?** It was this mystery that weighed on the mind of American astronomer Edwin Hubble as he sat peering through the eyepiece of his telescope in 1923, enshrouded in the darkness of the California night.

To settle the debate once and for all, Hubble had determined to establish a

reliable distance to the spiral nebulae and he had the right tool for the job. With a 254 cm (100 in) mirror, the Hooker Telescope at the Mount Wilson Observatory near Los Angeles was the most powerful in the world. Hubble turned its observational might on the **largest of the spiral nebulae – Andromeda** (also known as M31 – "M" for "Messier") – in the hope he might find a particular sort of star that he could use to calculate its distance.

In the 19th century, astronomers had figured out that there is an intrinsic link between the colour of a star and its temperature and brightness. If you can accurately identify the colour of a star, you can calculate how bright it would appear if you lived on a planet that orbited it. By knowing how bright it should appear and comparing that to its apparent brightness from Earth, you can figure out how far away it is. Known as **"Spectroscopic Parallax technique",** it is a terrifically accurate way to determine distance, but it only really works with relatively nearby stars. The further starlight has to travel, the more light-obscuring "stuff" (such as dust, which absorbs and reflects light) gets in its way. Eventually, the light that does make it through cannot be trusted to be telling the "truth" about the star it came from. Luckily, just over a decade before Hubble began his survey of Andromeda, a "computer" had found a solution.

In those days, computers were not rooms full of glowing valves and spinning data tapes – they were **women employed to study photographic plates and catalogue the brightness of stars.** In 1908, one of these "computers", the American Henrietta Swan Leavitt, discovered that astronomers like Hubble could exploit the properties of a particular kind of star to measure distances

Record breaker:
The Hooker Telescope was the largest telescope in the world when built in 1918. Hubble used it to discover how far away Andromeda is.

accurately over vast distances. Known as "Cepheid variables", these stars vary in brightness – throbbing from bright to dim like **cosmic Christmas tree lights.** Leavitt discovered that there was a link between how quickly they "throbbed" and their brightness (a Cepheid that takes ten days to go from bright to dim and back again will be brighter overall than one that takes seven days). If Hubble could find a Cepheid within Andromeda and measure its period of variation, he could determine its brightness and use that to calculate its distance.

Hubble spent several months in 1923 scanning Andromeda and making long photographic exposures in the hope of resolving an individual star. Pretty much all the stars that were bright enough to be spotted were so-called "novae" – white dwarf stars that suddenly brighten when **intense bursts of nuclear fusion** ignite on their surface (not to be confused with supernovae); these Hubble

dismissed by marking an "N" (for novae) next to the image.

On the night of 4 October 1923, Hubble made a 45-minute exposure that revealed three suspected novae, which he duly marked "N". But, two days later, he made another exposure and, when he compared it to the previous image, he realized that one of the "N"s had dimmed faster than it should. Over the following days he made enough observations to determine that the object was a Cepheid variable and he excitedly scribbled out the "N" and replaced it with "VAR!" (for variable). The new-found variable's period was 31.4 days – Hubble worked out its luminosity and calculated its distance as about a

million light-years – **well outside of the Milky Way and beyond the assumed limits of the Universe.**

By the end of 1924, Hubble had found 35 more variable stars in Andromeda (of which 12 were Cepheids). Far from being a small cloud on the fringes of the Milky Way, Andromeda was a whole other galaxy – made small only by the vast distance separating it from Earth. Later, observations of galaxies more distant than Andromeda revealed that they are **all rushing away from each other –** leading to the revelation that **the Universe is expanding and was born 13.8 billion years ago in the Big Bang** (Hubble was at the centre of that story, too).

Hubble crossed out his previous annotation of "N", for nova, and replaced it with a triumphant "VAR!", for variable.

EDWIN HUBBLE

Edwin Hubble was one of the most important astronomers of the 20th century. He created a classification system for galaxies; showed that there is something outside of the Milky Way; and discovered a link between a galaxy's redshift and its distance, which proved the Universe is expanding.

Away from the Milky Way:
This image taken by Hubble proved that there was something outside the confines of the Milky Way, which was then believed to be the full extent of the Universe.

HOW TO MEASURE DISTANCE IN SPACE

There are no tape measures in space, so astronomers had to come up with more inventive ways to measure distance. One method involves measuring the brightness of a star.

1 ### *Spectroscopy*
Using spectroscopy (a technique that splits light into its component colours), astronomers can obtain a spectral barcode, which tells them how bright it would appear if we were close to it (but not too close). They can then use this "absolute" brightness and compare it to its "apparent" brightness. Then they apply the inverse-square rule to estimate its distance.

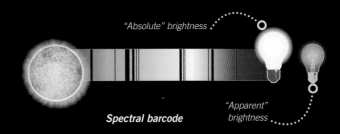

"Absolute" brightness

"Apparent" brightness

Spectral barcode

2 ### *The inverse-square rule*
As light travels through space, it spreads out in a sphere. But the number of photons remains the same. Photons twice as far from the light source are spread across four times the area (so are one quarter as bright). But for really distant stars, there is too much dust and gas in the way to get an accurate spectrum. Hubble used a star called a Cepheid variable to get around this.

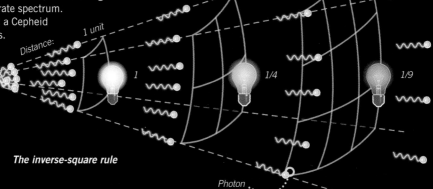

3 units

2 units

1 unit

Distance:

1

1/4

1/9

The inverse-square rule

Photon

3 ### *Special stars*
Cepheid variables swell and contract – pulsing from bright to dim and back to bright again over a measurable period. The period is determined by the star's luminosity – the amount of light the star produces. By studying a Cepheid's period, astronomers can determine its absolute brightness and then use the inverse-square rule to measure its distance.

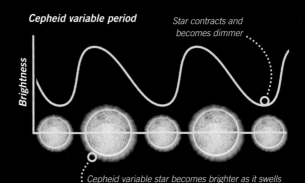

Cepheid variable period

Star contracts and becomes dimmer

Brightness

Cepheid variable star becomes brighter as it swells

Star cluster:
This star cluster is R136. It can be found in a colossal star-forming nebula called the 30 Doradus Nebula, which is the largest and most prolific stellar nursery in the Milky Way.

EXPANDING
UNIVERSE

WE USED TO BELIEVE that stars were eternal and the Universe was infinite and immutable, but we now know that this is not the case. From vast clouds of cosmic gas, stars condense, ignite, and burn themselves to death. **Even the Universe had a moment of birth and one day, it, too, will die.** But surely, as long as there is a Universe to house them, there will be stars?

Maybe not. **Could it be that, one day, mankind's distant descendants will gaze at the night sky and see a starless carpet of perfect black?**

A study from 2012 suggests that the best days of the Universe's star formation are long behind it, and that most of the stars left **are now creeping into old age.**

In the most comprehensive study of its kind, scientists used three massive telescopes to look at star-forming galaxies from four billion to 11 billion years ago. They used the data to chart the history of star formation in the Universe, and found that, **in its early days, the Universe was far more prolific in its star-forming activities than it has been in the last few billion years.** In fact, the researchers concluded that **95 per cent of the Universe's stars have already been formed.**

All stars start off by using hydrogen to fuel their nuclear furnaces and, as the stars age, this hydrogen gets fused into increasingly heavier elements. It seems that there is **just not enough hydrogen left in galaxies to keep forming new stars.**

That is not to say that there are not billions of stars yet to be made: a huge drop from a colossal figure is still a very large number indeed. So there will be stars decorating the heavens for some time to come, but **it may be that anything that lies beyond our own galaxy will not be visible from Earth.** According to one theory about the fate of the Universe, all those galaxies that make the heavens a more interesting place could be expelled from the night sky as the **expanding Universe carries them from sight.**

A BLACK SKY?

In billions of years' time, the light from stars within distant galaxies will be unable to outrun the Universe's acceleration, and we will know of nothing beyond the environment of the Milky Way. Although the nearest galaxies will remain gravitationally bound to the Milky Way, and so remain visible, everything else that makes up the Universe will recede from sight. On the plus side, Earth will probably be long gone by then...

HOW THE UNIVERSE
WILL BANISH GALAXIES

Until the 20th century, it was believed that the Universe was eternal, unchanging, and infinite, but there was a problem with this idea – it just did not match the evidence...

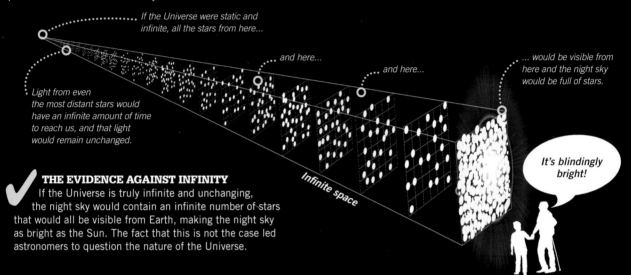

If the Universe were static and infinite, all the stars from here...

and here...

and here...

... would be visible from here and the night sky would be full of stars.

Light from even the most distant stars would have an infinite amount of time to reach us, and that light would remain unchanged.

Infinite space

It's blindingly bright!

✓ THE EVIDENCE AGAINST INFINITY
If the Universe is truly infinite and unchanging, the night sky would contain an infinite number of stars that would all be visible from Earth, making the night sky as bright as the Sun. The fact that this is not the case led astronomers to question the nature of the Universe.

✓ THE QUICK AND THE RED
Astronomers noticed that the light from distant galaxies was redder than it should have been, with the light appearing redder in more distant stars. Light is part of the electromagnetic spectrum and therefore has a wavelength. Light at the red end of the spectrum has a longer wavelength, and is further away, than light at the blue end.

Infrared   **Visible light**   **Ultraviolet**

← **Redshift** | **Blueshift** →

As the galaxy moves away, its speed increases – the further away it is, the faster it gets.

The further away the galaxy, the longer the wavelength.

I like red!

REDSHIFT
The light emitted was being stretched into the red end of the spectrum (called redshift). The answer must be that the galaxies are actually moving.

✓ THE GREED FOR SPEED
The discovery that stars and galaxies are all rushing away from each other led to the revelation that (far from being static) the Universe is actually expanding. If it is growing, it must have had a birth (which we now call the "Big Bang"). We cannot see every star that exists because the Universe has not been around for long enough for the light from the most distant stars to reach us.

According to Einstein's theory of relativity, photons can move at a maximum of 300,000 km (186,411 miles) per second.

Catch me if you can!

LIGHT SPEED
Light is made up of packets of energy called photons. As they have a maximum speed, light from distant galaxies takes billions of years to reach us.

STANDING STILL AT THE SPEED OF LIGHT

We now know that the rate at which the Universe is expanding is accelerating and all those stars and galaxies rushing away from us are getting faster and faster. It is possible that they could eventually appear to be moving away from us faster than the speed of light. But Einstein tells us that nothing can travel faster than the speed of light, so how can this be so?

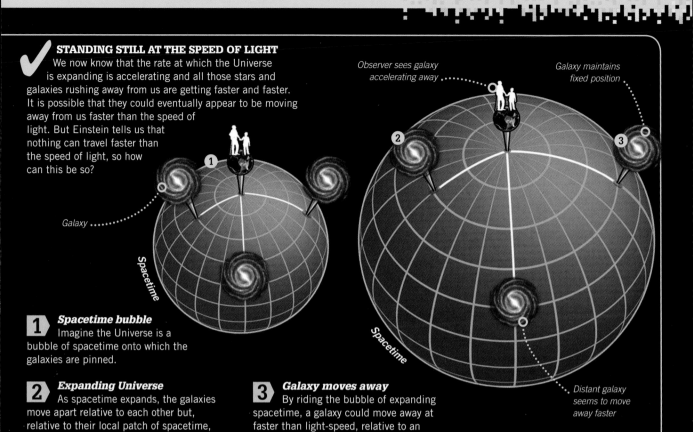

Observer sees galaxy accelerating away

Galaxy maintains fixed position

Galaxy

Spacetime

Spacetime

1 Spacetime bubble
Imagine the Universe is a bubble of spacetime onto which the galaxies are pinned.

2 Expanding Universe
As spacetime expands, the galaxies move apart relative to each other but, relative to their local patch of spacetime, they have not really moved at all.

3 Galaxy moves away
By riding the bubble of expanding spacetime, a galaxy could move away at faster than light-speed, relative to an observer on Earth.

Distant galaxy seems to move away faster

HOW TO MAKE A GALAXY DISAPPEAR

If the space between Earth and the receding galaxy expands faster than the photons of light emitted by stars within it can travel, that light will never reach Earth. Imagine, if you will, the photons of light are on a spacetime travellator.

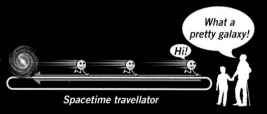

What a pretty galaxy!

Hi!

Spacetime travellator

TRAVELLATOR
As the Universe's expansion accelerates, the travellator gets faster and faster. Eventually, the travellator is moving faster than the photons can run along it, and they never reach the other end – meaning their light never reaches us.

Urgh!

Where did it go?

Spacetime travellator expands and accelerates

ALBERT EINSTEIN

In his theory of special relativity, German-born scientist Albert Einstein revealed that light had a maximum speed limit. As a result, if space is expanding faster than the speed of light, the source of that light will vanish from sight.

WELCOME TO THE MULTIVERSE

IS OUR UNIVERSE just one bubble of existence in an infinite **multiverse**? The latest survey of the **cosmic microwave background** (CMB, also known as the radiation after-glow of the Big Bang) by the European Space Agency's (ESA's) Planck space telescope seemed to uphold an idea called **"cosmic inflation".** Cosmic inflation is a sort of "injection" of energy that **caused the Universe to expand exponentially just moments after the Big Bang,** when it was still smaller than an atom. This rapid inflation is seen by many as being the only explanation for the apparent even spread of energy in the early Universe (by inflating the teeny tiny Universe before it got the chance to spread slowly and get all "lumpy"). But a potential consequence of cosmic inflation is that, while most of the Universe slowed down, tiny pockets could have continued their exponential inflation – **creating off-shoot "bubble" universes.**

Another tantalising hint of the **existence of other universes** can be found in Planck's cosmic microwave background survey. There is a mysterious cold spot (pictured below, far right) that, some have suggested, **could be the "imprint" left behind by another Universe before it separated from our own.** Although there is nothing in current cosmological theory that explicitly rules out the existence of other universes, there is no hard

The universes pop off once they are formed.

evidence supporting the idea, either. But, **it is fun to imagine that it might be possible.**

One of the most commonly asked questions of Big Bang theorists is **"what came before the Big Bang?"** The standard answer is that there was nothing at all. Normally, we are quite comfortable with the idea of "nothing". We are conditioned by experience to think of nothing as being an absence of something within a given area, but if space itself was created in the Big Bang, there cannot be "nothing" because there is nowhere to put the "something" that does not exist. **Asking what came before the Big Bang is equally meaningless because "time" was created along with space** – you cannot have a "before" because time did not exist. For a species that experiences the world by interacting with time and space, that is **a slightly uncomfortable, brain-blending concept.**

Luckily (depending on your point of view) there are physicists who believe that, far from being the beginning of all things, the Big Bang was just the moment our Universe burst from the womb of a parent Universe – **just one offspring of a much larger multiverse.**

The idea that our Universe is just one of countless others might seem (at best) incredible and (at worst) delusional, but remember this: we once thought our planet was unique; then we thought our Solar System was unique and, after that, we thought our galaxy was unique – **is it such a stretch to imagine that our Universe is not unique, as we like to believe?** One of the problems with our Universe being "the" Universe is that it seems a little too perfect. It is a Universe where the laws of physics are perfectly tuned for the creation of stars, galaxies, planets, and life – if just one aspect of those laws were different, then the Universe as we know it would not exist. It is

Each universe may have different rules and outcomes.

ARE WE PART OF A LARGER MULTIVERSE?

COLD SPOT

When scientists analysed the Planck CMB data, they noticed that a region of sky near the constellation Eridanus was colder than the surrounding region. This "cold spot" is a highly peculiar anomaly that might be an imprint mark left behind by another universe.

Cold spot

the same problem we once had with our planet.

Looked at in isolation, Earth seems to have been perfectly "designed" for the creation of life – just the right distance from just the right sort of star with just the right atmosphere and just the right sort of magnetic field (and so on). Of course, we now know that there are countless other planets out there where conditions are not perfect and where life does not exist. **We were just the winners in the planetary lottery.** The multiverse would solve the

problem of our "perfect" Universe in the same way. Just as Earth won the planetary lottery, our Universe won the cosmological lottery. It seems "perfect" because the conditions within it allowed us to evolve and marvel at its perfection. **But there are countless other universes where conditions were not just right.** You can compare it to a game of cards. If you were allowed to pull just one card from the deck, the chances of pulling out the card you were looking for is quite small, but, if you were allowed to go through the whole

deck, your card's discovery becomes inevitable. The same applies to the multiverse: with infinite permutations of the laws of physics available, it is inevitable that one would be perfect for life. In many ways, a multiverse is a more comfortable concept to get to grips with than a perfect Universe born from the void.

Of course, eventually you have to ask where the first of these ancestor universes came from, and **you are right back where you started!**

CHILDREN OF THE BLACK HOLE

Another theory is that our Universe was born within a black hole and that black holes within our cosmos are creating universes of their own.

1 *Black hole*
This is a black hole. She is quite happy munching her way through all the light and matter that strays too close to her irresistible gravitational pull. At her heart is a tiny ball of concentrated matter called a singularity, which gets increasingly compact as it gains mass, until it reaches near-infinite density.

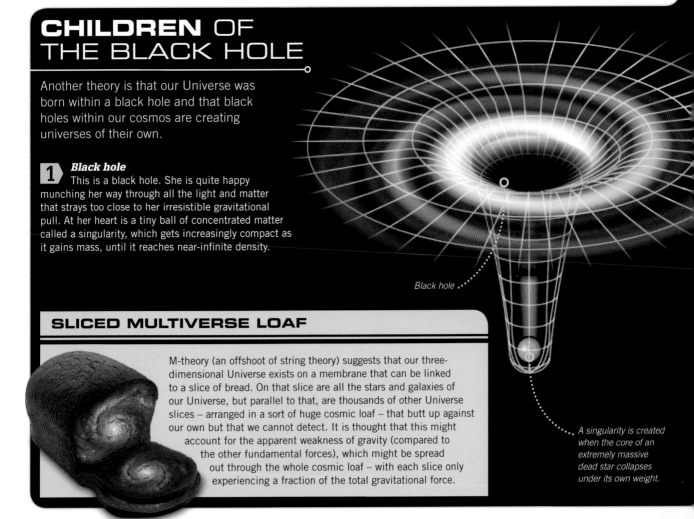

Black hole

SLICED MULTIVERSE LOAF

M-theory (an offshoot of string theory) suggests that our three-dimensional Universe exists on a membrane that can be linked to a slice of bread. On that slice are all the stars and galaxies of our Universe, but parallel to that, are thousands of other Universe slices – arranged in a sort of huge cosmic loaf – that butt up against our own but that we cannot detect. It is thought that this might account for the apparent weakness of gravity (compared to the other fundamental forces), which might be spread out through the whole cosmic loaf – with each slice only experiencing a fraction of the total gravitational force.

A singularity is created when the core of an extremely massive dead star collapses under its own weight.

2 **Singularity bounces back**
At this point, standard theory suggests that space and time become so heavily distorted at the singularity that time stops. But one theory says that the singularity "bounces back" and punches a hole in spacetime (the fabric of the Universe).

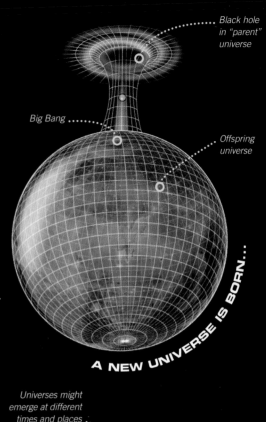

Black hole in "parent" universe

Big Bang

Offspring universe

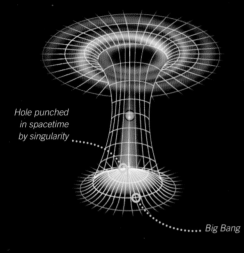

Hole punched in spacetime by singularity

Big Bang

3 **Big Bang**
Here, the singularity begins to expand – creating a "Big Bang" from which a new universe is born, where the laws of physics might be slightly different from those of its parent universe.

A NEW UNIVERSE IS BORN...

Black hole collapses in "parent" universe

Universes might emerge at different times and places

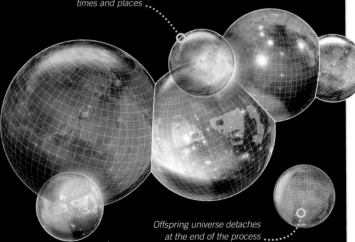

Offspring universe detaches at the end of the process

4 **Time stops in singularity**
Back in the parent universe, just as time starts in the new universe, time stops at the singularity. Eventually, the original black hole collapses – severing the umbilical cord to the offspring universe.

5 **Multiverse**
There might be an infinite chain of universes, but only a few in which the laws of physics are conducive to life.

WE'RE ALL DOOMED!

MORE THAN 400 MILLION YEARS AGO, Earth was a very different place from today. The climate, encouraged by excessive levels of greenhouse gases, was hot enough to ensure no water was locked away at the poles. **Sea levels were a great deal higher than today** and, in the balmy waters, sea life had exploded – dominating life on Earth.

Then this all changed. Inexplicably, the climate cooled, glaciers formed at the poles, sea levels plummeted, and more than 85 per cent of Earth's species died out. The Late Ordovician mass extinction, as it has come to be known, was **one of the most catastrophic extinction events in the planet's history.** But what caused the climate to take such a dramatic U-turn?

One explanation is that Earth was the **unwilling recipient of a massive dose of gamma radiation gifted to us by a distant star dying a violent death.** Gamma ray bursts (GRBs) are the most powerful cosmic explosions we know of. When a massive star explodes in a supernova explosion, sometimes, as if in a fit of raw fury, **the star will spew intense beams of deadly gamma radiation into the cosmos.**

If Earth was at the receiving end of such an outburst all those millennia ago, gamma radiation would have wreaked havoc on the planet. It would have **destroyed the protective ozone layer, and blanketed the planet in a suffocating layer of smog** that would have blocked sunlight and sent the climate into a tailspin, resulting in the **death of more than three-quarters of life on Earth.** Well, get your best apocalypse trousers on, because if some scientists are to be believed, **Earth could be on the receiving end of another dose of gamma rays soon.**

Eight thousand light-years away, in the constellation of Sagittarius (the archer), a dying star could have us locked in its sights. WR 104 is a Wolf-Rayet binary star system composed of **two truly massive stars engaged in an orbiting death dance.** Due to their size (equivalent to as many as 20 Suns each), both stars are living on borrowed time and will

> **WOLF-RAYET STARS ARE THE MOST MASSIVE AND BRIGHTEST STARS KNOWN**

soon **die the sort of violent death that only truly colossal stars can.** But one is a very special sort of star called a Wolf-Rayet star.

When this star dies, not content with an understated supernova, its core will collapse to form a black hole that, through a fierce collusion of forces, **will vent two beams of gamma radiation along the star's poles and out into space – possibly towards us.**

Since WR 104 was discovered in 1998, arguments have swung back and forth as to whether it has us in its sights. Now two astronomers at the Keck Observatory in Hawaii, USA, have suggested that we could be better aligned with the so-called "Death Star's" poles than we might find comfortable.

As the two stars orbit each other, they vent huge quantities of material that spread out to create a spiral pattern. When we look at the spiral from Earth, we appear to be seeing it face-on – **meaning the star's poles could be pointing at our little blue planet.**

Although the star could go boom at any time in the next 500,000 years, the slightest misalignment of even a few degrees would see the beam sail by harmlessly. So we might not be so doomed after all.

MEET THE REAL "DEATH STAR"...

WR 104 consists of two stars, both of which are extremely massive and counting down to their imminent demise. But it is the Wolf-Rayet star that could create the potentially lethal gamma ray burst.

1 ***Crushed core***
When a star runs out of fuel, nuclear reactions shut down in its core. For a star as massive as a Wolf-Rayet, this causes the star to explode as a supernova, and the core to collapse catastrophically under the weight of its own gravity.

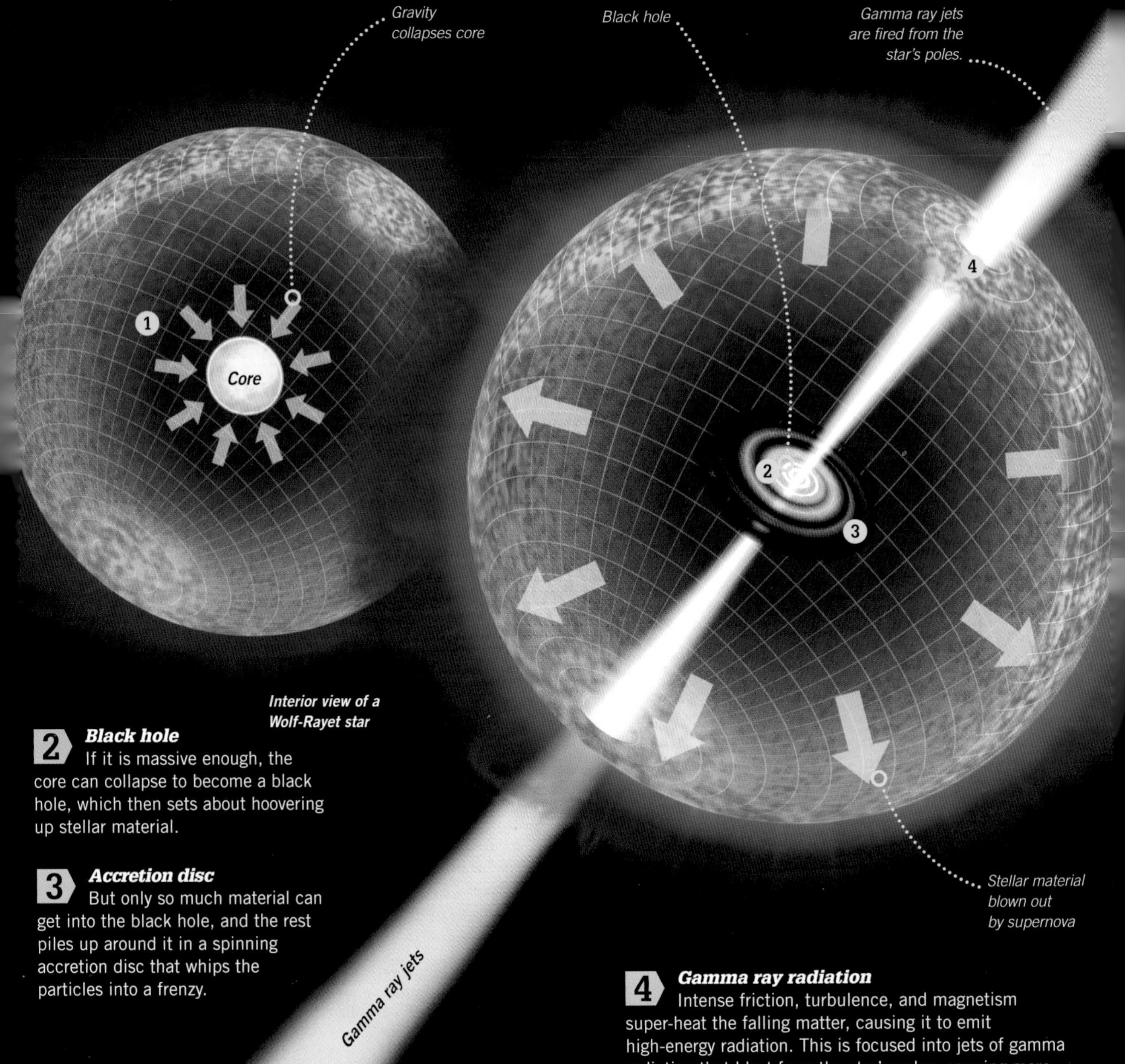

Gravity collapses core

Black hole

Gamma ray jets are fired from the star's poles.

Core

Interior view of a Wolf-Rayet star

Gamma ray jets

Stellar material blown out by supernova

2 ***Black hole***
If it is massive enough, the core can collapse to become a black hole, which then sets about hoovering up stellar material.

3 ***Accretion disc***
But only so much material can get into the black hole, and the rest piles up around it in a spinning accretion disc that whips the particles into a frenzy.

4 ***Gamma ray radiation***
Intense friction, turbulence, and magnetism super-heat the falling matter, causing it to emit high-energy radiation. This is focused into jets of gamma radiation that blast from the star's poles, carrying more energy than our Sun will put out in its entire lifetime.

EARTH IN ITS SIGHTS...

A blast of gamma ray radiation would have a catastrophic effect on the atmosphere and life on Earth. Though we can't do much to stop it from happening, we might just be able to see if it's coming our way...

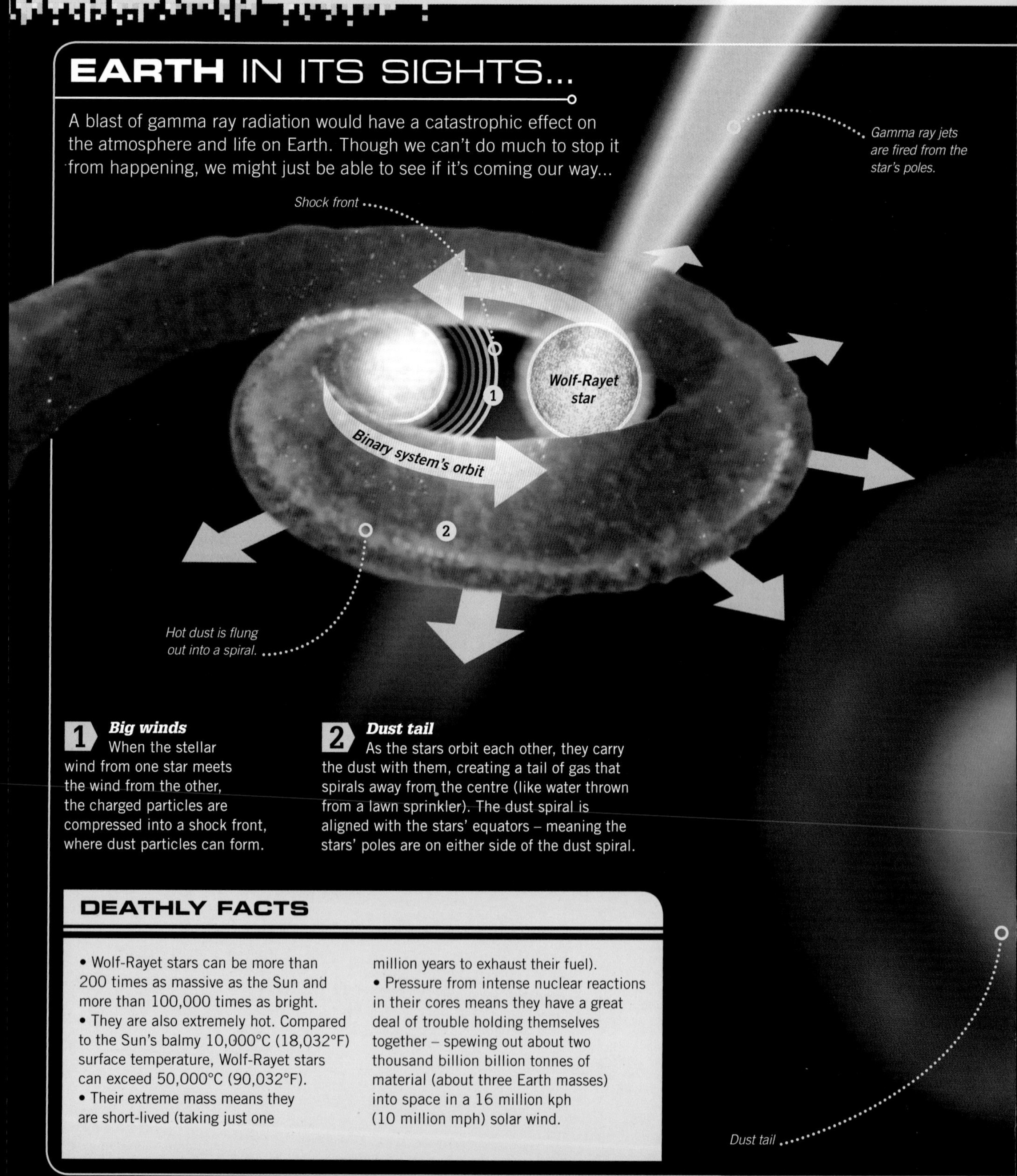

Gamma ray jets are fired from the star's poles.

Shock front

Wolf-Rayet star

1

Binary system's orbit

2

Hot dust is flung out into a spiral.

1 Big winds
When the stellar wind from one star meets the wind from the other, the charged particles are compressed into a shock front, where dust particles can form.

2 Dust tail
As the stars orbit each other, they carry the dust with them, creating a tail of gas that spirals away from the centre (like water thrown from a lawn sprinkler). The dust spiral is aligned with the stars' equators – meaning the stars' poles are on either side of the dust spiral.

DEATHLY FACTS

• Wolf-Rayet stars can be more than 200 times as massive as the Sun and more than 100,000 times as bright.
• They are also extremely hot. Compared to the Sun's balmy 10,000°C (18,032°F) surface temperature, Wolf-Rayet stars can exceed 50,000°C (90,032°F).
• Their extreme mass means they are short-lived (taking just one million years to exhaust their fuel).
• Pressure from intense nuclear reactions in their cores means they have a great deal of trouble holding themselves together – spewing out about two thousand billion billion tonnes of material (about three Earth masses) into space in a 16 million kph (10 million mph) solar wind.

Dust tail

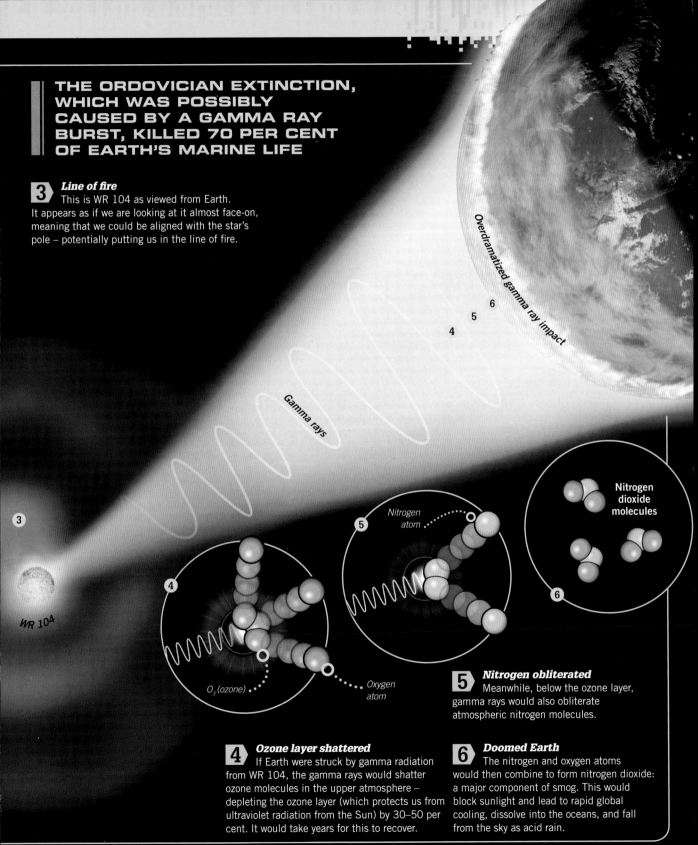

THE ORDOVICIAN EXTINCTION, WHICH WAS POSSIBLY CAUSED BY A GAMMA RAY BURST, KILLED 70 PER CENT OF EARTH'S MARINE LIFE

3 *Line of fire*
This is WR 104 as viewed from Earth.
It appears as if we are looking at it almost face-on,
meaning that we could be aligned with the star's
pole – potentially putting us in the line of fire.

Overdramatized gamma ray impact

6
5
4

Gamma rays

**Nitrogen
dioxide
molecules**

3

WR 104

*Nitrogen
atom*

5

Nitrogen obliterated
Meanwhile, below the ozone layer,
gamma rays would also obliterate
atmospheric nitrogen molecules.

6

4

O_3 *(ozone)*

*Oxygen
atom*

4 *Ozone layer shattered*
If Earth were struck by gamma radiation
from WR 104, the gamma rays would shatter
ozone molecules in the upper atmosphere –
depleting the ozone layer (which protects us from
ultraviolet radiation from the Sun) by 30–50 per
cent. It would take years for this to recover.

6 *Doomed Earth*
The nitrogen and oxygen atoms
would then combine to form nitrogen dioxide:
a major component of smog. This would
block sunlight and lead to rapid global
cooling, dissolve into the oceans, and fall
from the sky as acid rain.

CATCH UP WITH THE STELLAR SPEED DEMONS

A SHOOTING STAR JUST DOES NOT LIVE UP TO the beauty of its name. It promises a flaming stellar projectile fired from the cannon of the gods, but in reality, a shooting star is less colossal spherical inferno and more **a speck of cosmic dust that lost a battle with friction.** Luckily, the Universe (that great purveyor of baffling wonderment) **has some real shooting stars up its sleeve.**

Sometimes called rogue, runaway, or even hypervelocity stars, these are stars that have been liberated from the gravitational bonds of the galaxy, and **set free to travel the cosmos at almost unimaginable speeds.**

Since the first one was sighted in 2005, dozens of these intergalactic speed demons have been found **careering around the cosmos like stellar boy racers.**

Most of the Milky Way's hundred billion or so stars orbit the galactic centre at a relatively pedestrian 640,000 kph (400,000 mph) or so, but **hypervelocity stars can be travelling at 3,200,000 kph (2,000,000 mph) and some might be streaking along at many times that speed.** Hypervelocity stars move so fast that **they can exceed the escape velocity of the galaxy** (the speed needed to overcome the pull of gravity). This is no mean feat because the object they are escaping is a supermassive black hole that weighs in at four million times the mass of our Sun – which is a lot of gravity to overcome.

Why do these stars go "rogue" in the first place? When their existence was first proposed, astronomers believed their discovery would confirm the then-theoretical existence of a black hole at a galaxy's heart because **only a supermassive black hole could provide the gravitational "kick"** needed to accelerate a massive star to hypervelocity speeds.

Close to the galactic centre, gravity is so extreme that stars in this region are whipped into super-fast orbits that see them **whizzing along their elliptical motorways at more than a million kilometres per hour.** It is a delicate balancing act that, if disturbed, could see the star being sucked to its doom or flung into the dark expanse of intergalactic space.

EVEN AT HYPERVELOCITIES, A STAR WOULD TAKE ABOUT 10,000,000 YEARS TO TRAVEL FROM THE CENTRE OF THE MILKY WAY TO ITS EDGE

ALPHA CAM

One of the fastest hypervelocity stars yet discovered is called Alpha Camelopardalis, or Alpha Cam. In this image, taken by NASA's WISE space telescope, the red splodge is a band of glowing gases heated up by the shock wave created by the star as it streaks through the cosmos. Its speed has been estimated at between 2.4 and 15 million kph (1.5 and 9.4 million mph) – at that speed it would take just over a second to travel from London to New York.

INSIDE **THE GALAXY**

At the centre of the Milky Way there is a colossal black hole with the mass of four million suns. More distant stars, like our Sun, orbit the galactic centre at about 720,000 kph (450,000 mph), but a closer star can orbit at millions of kilometres per hour. It is thought that for even such a speedy star to be kicked out of the galaxy, it must be part of a binary, or multiple, star system.

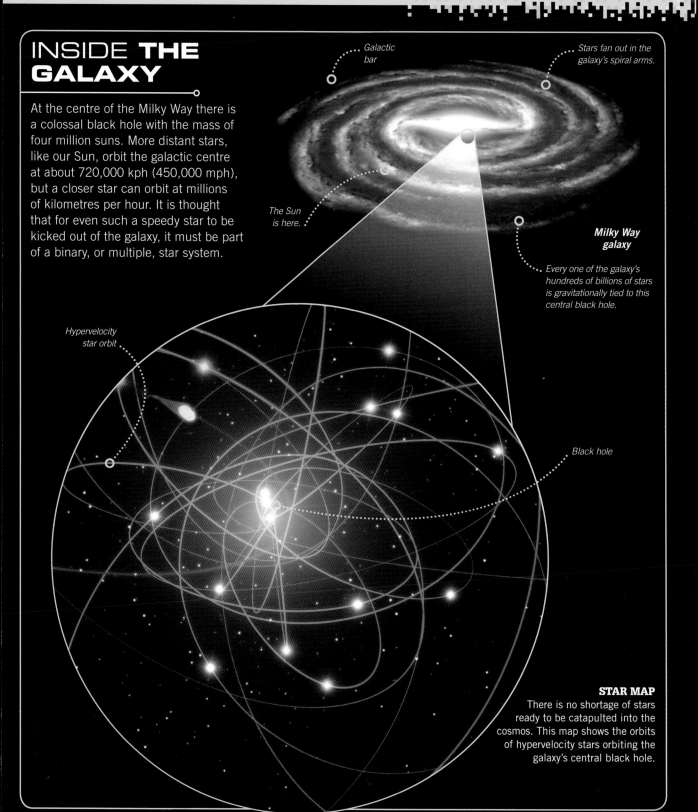

Galactic bar

Stars fan out in the galaxy's spiral arms.

The Sun is here.

Milky Way galaxy

Every one of the galaxy's hundreds of billions of stars is gravitationally tied to this central black hole.

Hypervelocity star orbit

Black hole

STAR MAP
There is no shortage of stars ready to be catapulted into the cosmos. This map shows the orbits of hypervelocity stars orbiting the galaxy's central black hole.

HOW HYPERVELOCITY STARS ARE THROWN OUT OF THE GALAXY

The existence of hypervelocity stars was first proposed in 1988, but the first was not discovered until 2005. We now know that these rogue stars travel so fast that they can escape the gravitational confines of the galaxy that bore them and travel out into intergalactic space. Here's how...

ASTRONOMERS THINK THAT ONE HYPERVELOCITY STAR IS BOOTED OUT OF THE GALAXY EVERY 10,000 YEARS

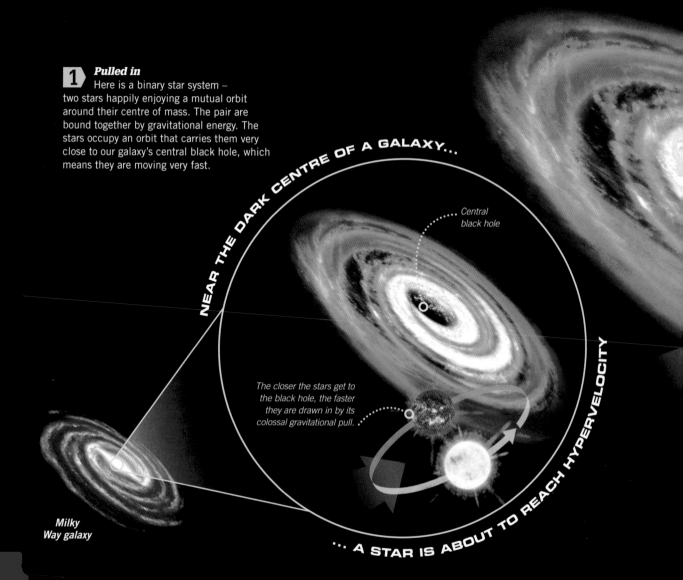

1 Pulled in
Here is a binary star system – two stars happily enjoying a mutual orbit around their centre of mass. The pair are bound together by gravitational energy. The stars occupy an orbit that carries them very close to our galaxy's central black hole, which means they are moving very fast.

NEAR THE DARK CENTRE OF A GALAXY...

Central black hole

The closer the stars get to the black hole, the faster they are drawn in by its colossal gravitational pull.

... A STAR IS ABOUT TO REACH HYPERVELOCITY

Milky Way galaxy

2 *Torn apart*

But if they stray too close, the black hole's colossal gravitational energy can overwhelm the energy that binds the stars together, and tear them apart. One star is captured, and falls into the black hole.

3 *Speed boost*

Before they are torn apart, the two stars were orbiting at great speed. When the pair's gravitational bond is broken, all the energy and angular momentum within the system is transferred to the remaining star – giving it a massive boost, and firing it outwards.

Star gets sucked into the black hole.

4 *Hypervelocity!*

The star's initial orbital speed, combined with the extra energy and momentum, accelerates the star to up to 15 million kph (9.4 million mph) and flings it out of the galaxy. It wanders the Universe as a stellar speed demon.

WOOSH!

The star is flung out and into space.

Remaining star is sent outwards.

SUPERNOVA SLINGSHOT

But the black-hole-slingshot model does not fit for all hypervelocity stars. Some seem to have been "spat out" from more distant regions of the galaxy, and many of them contain chemicals that could only have come from a supernova explosion. One explanation for this is that these stars were once part of an extremely rapidly rotating binary pair of stars. When one of the pair exploded as a supernova, the gravitational release combined with the extra "shove" of the explosion gave the remaining star enough speed to escape the galaxy.

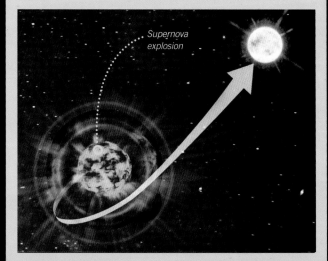

Supernova explosion

MEET THE SMELLY DWARF

IN A UNIVERSE POPULATED BY THE BIZARRE and unusual, it takes a special talent to be singled out as a **space oddity,** but if there is one celestial object that deserves this moniker, it is the **lowly brown dwarf.**

Stuck in a strange no-man's land between stars and planets, and accused of being dull, **smelly** (their atmosphere is rich with **eggy hydrogen sulphide** and **urine-smelling ammonia**), and underachieving loners, brown dwarfs are one of the Universe's most maligned objects – a sort of cosmic hobo if you will.

Formed from the collapse of clouds of gas and dust, brown dwarfs start their lives full of **the promise of stardom.** But they never manage to gather enough mass to ignite full-blooded hydrogen fusion in their cores and, instead of becoming blazing stars surrounded by supplicant planets, they resemble enormous Jupiter-like planets – **doomed to billions of years of cold obscurity.**

Brown dwarfs are so unlike stars that their existence was only actually confirmed in 1994 after decades of speculation. Although they begin life as hot, dense almost-stars, without a functioning nuclear furnace in their cores, brown dwarfs haemorrhage heat into the frigid vacuum of space and quickly become **too cold to be seen by conventional telescopes.** Luckily, although they can be feebly cold by stellar standards, they are warmer than the space that surrounds them, which makes them visible to infrared telescopes. But this does not make them easy to find.

A young brown dwarf can be hot enough to be mistaken for the smallest stars and an old brown dwarf can be cold enough to be mistaken for a Jupiter-like gas giant. So instead of thinking of brown dwarfs as being the hobos of the Universe, **perhaps we should imagine them as cosmic double agents –** surely they deserve that much?

THE BROWN DWARF:
A STELLAR DROPOUT

The early careers of brown dwarfs and stars are very similar – both are formed from the gravitational collapse of clouds of interstellar gas and dust – but, somewhere along the line, a brown dwarf drops out of the stellar university and lives the rest of its life labelled as a "failed star".

1 *Gas and dust cloud*
Here is a cloud of interstellar gas and dust. It is mostly made up of hydrogen and helium, with small amounts of deuterium (also known as "heavy hydrogen") and lithium. The centre is slightly denser than its surroundings, and this acts as a sort of seed around which a star can grow.

2 *Protostar*
The extra gravitational pull of the seed draws in material from the cloud until a young protostar is formed.

Cloud material is drawn into the centre

3 *Deuterium fusion*
The protostar gains mass and contracts under its own weight. As it gets more dense, its core heats up until it is hot enough to fuse its deuterium. The energy created by deuterium fusion heats the protostar to several million degrees and stops it shrinking.

4 *Empty fuel tank*
But here its dreams of stardom end. It soon runs out of deuterium fuel and it has not gained enough mass to keep the fusion going (it takes a lot more energy to fuse hydrogen).

5 *Cooling off*
With no new energy being created in its core, it starts to shrink and cool.

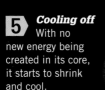

Unused gas and dust drift away

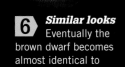

6 *Similar looks*
Eventually the brown dwarf becomes almost identical to a gas giant planet.

7 *Undetectable*
In time, it will cool completely and become almost undetectable.

FAILED STAR?
OR OVERACHIEVING PLANET?

Gas giant planets, brown dwarfs, and stars are all made of pretty much the same ingredients: about 90 per cent hydrogen and 10 per cent helium. So is it fair to call a brown dwarf a failed star, or can we call it an overachieving planet instead?

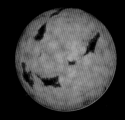

HEAT: THE STAR QUALITY

Here are three objects that are made of the same stuff and are about the same size. Jupiter is the largest planet in our Solar System, Gliese 229B is a fairly typical brown dwarf, and OGLE-TR-122B (yes, that is its name) is the smallest star yet discovered.

JUPITER, GAS GIANT PLANET	GLIESE 229B, BROWN DWARF STAR	OGLE-TR-122B, SMALL STAR
Mass: 1 Jupiter mass	**Mass:** 40 Jupiter masses	**Mass:** 96 Jupiter masses
Temperature: -145°C (-229°F)	**Temperature:** 730°C (1,346°F)	**Temperature:** 2,500°C (4,500°F)
Despite its size, Jupiter is not particularly dense, so it does not have enough mass to force the atoms that make it up to fuse together. This means it is pretty darn cold.	Gliese 229B is far more massive than Jupiter. The gravitational force of the extra mass is enough to make deuterium atoms fuse in its core – generating heat for a short time.	OGLE-TR-122B is more than twice as massive as Gliese 229B. This creates enough core pressure for hydrogen fusion to take place – creating heat for many millions of years.

BUILT LIKE A STAR
(AND MAYBE A PLANET)

It was assumed that Jupiter-like gas giants formed more like planets than stars. In recent years, gas giants have been discovered outside our Solar System that are too young to have been "built" like planets. So, if some planets are built like stars, we might be able to call brown dwarfs planetary overachievers.

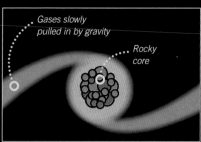

Gases slowly pulled in by gravity

Rocky core

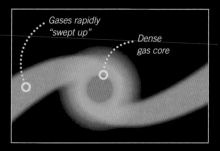

Gases rapidly "swept up"

Dense gas core

IT IS THOUGHT THAT THERE MIGHT BE AS MANY AS 30–100 BILLION BROWN DWARFS IN OUR GALAXY ALONE

OLD THEORY: ACCRETION MODEL
The gas giant starts off as a baby rocky planet but, over millions of years, its rocky core becomes wrapped in gases left over from the formation of their star. This requires the planet to have a rocky core, which brown dwarfs do not. Under this model, brown dwarfs remain "failed stars".

NEW THEORY: COLLAPSE MODEL
Gas giants form from the collapse of the gas disc of a still-growing star. In place of a rocky core, the planet grows around a seed of dense gas. This is similar to how stars and brown dwarfs are formed. Under this model we could call brown dwarfs "overachieving planets".

MERCURY'S
SECRETS

WHEN ANCIENT ASTRONOMERS observed the tiny planet that lives on the doorstep of the Sun, they saw how quickly it seemed to shoot across the heavens. So the Romans named it after the god Mercury – **the swift messenger.** When the scientists of the 21st century decided to send the first spacecraft to orbit Mercury, they saw no reason to break the tradition, **so they named the craft** *MESSENGER*.

MESSENGER was launched in 2004 and, since moving into orbit in early 2011, **the spacecraft has captured nearly 100,000 images** and revealed new information about Mercury, which, even many thousands of years since its discovery, has always been one of the **least understood planets in our Solar System.**

Mapping Mercury:
All we had were close-up pictures of one half of Mercury's surface until *MESSENGER* made its first fly-by of the planet in 2008. *MESSENGER* has now mapped more than 99 per cent of the planet.

THE SUN-FACING SIDE OF *MESSENGER*'S SHIELD REACHES TEMPERATURES OF 360°C (700°F)

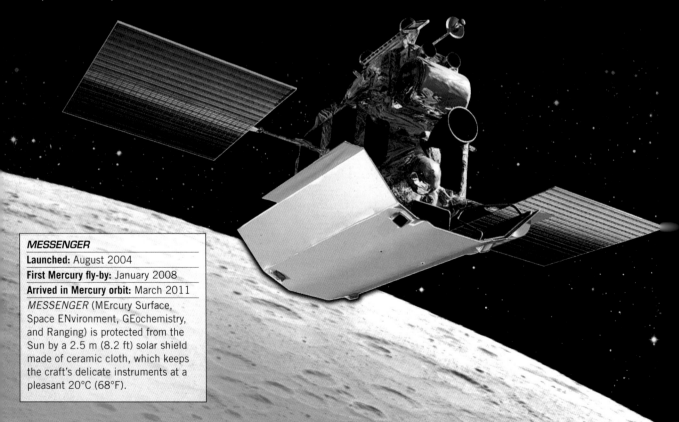

MESSENGER

Launched: August 2004

First Mercury fly-by: January 2008

Arrived in Mercury orbit: March 2011

MESSENGER (MErcury Surface, Space ENvironment, GEochemistry, and Ranging) is protected from the Sun by a 2.5 m (8.2 ft) solar shield made of ceramic cloth, which keeps the craft's delicate instruments at a pleasant 20°C (68°F).

The reason we know so little about Mercury is simple – **it is uncomfortably close to the Sun.** It is difficult to study from afar, because telescopes like Hubble are blinded by the Sun's glare and any probe sent to Mercury must contend with temperatures that swing from a searing 360°C (680°F) to a frigid -160°C (-256°F). *Mariner 10* was previously the only craft to make the journey – it snapped a handful of images of the planet as it whizzed past in 1974.

Thirty years after *Mariner 10*, *MESSENGER* was sent to shed light on this most illuminated of planets. After a year in orbit, *MESSENGER*'s findings are **revealing Mercury to be a most surprising and complex little world.**

Data from *MESSENGER* has allowed scientists to build the first precise model of Mercury's gravity field, which, combined with topographic data, has shed light on the planet's internal structure. It has revealed that Mercury's enormous core (relative to the planet's size) is **unlike anything in the Solar System.** It seems that the complex core, which accounts for 85 per cent of the planet's diameter, consists of a solid iron core, sitting within a ball of molten iron and encased by a sphere of solid iron sulphide (sort of like a giant iron Ferrero Rocher chocolate... or Ferrous Rocher).

Perhaps most surprisingly, the craft also seems to have found that, despite the planet's proximity to its fiery neighbour, **there may be water hiding at Mercury's poles.**

MERCURY **INSIDE OUT**

Mercury is only slightly bigger than Earth's Moon and is the closest planet to the Sun. During its long 58-hour day (a year lasts just 88 days) the surface is superheated to 430°C (800°F), while at night, temperatures drop as low as -180°C (-300°F). Despite its proximity to the Sun, Mercury's reflective surface and lack of atmosphere mean it isn't the Solar System's hottest planet – nearby Venus has that dubious honour.

Crust

Mantle

Solid iron sulphide shell around core

Liquid iron outer core

Magnetic field

-180°C (-300°F)

Night side

WATER AT THE POLES?
In the 1990s, astronomers discovered patches that reflected radar rays sent from Earth near Mercury's poles. At the time, it was suggested that water ice could be hiding in the cold, permanently shadowed craters. Scientists have compared these bright radar patches with new images taken by *MESSENGER* and they seem to match up perfectly – suggesting that, despite Mercury's close proximity to the Sun, water may be hiding at its poles.

THE SURFACE

Before *MESSENGER*, many scientists believed that the planet had cooled off quickly after its formation and has been geologically dead for billions of years. *MESSENGER's* findings seem to indicate that geological processes continued for some time after the planet's formation.

Moving on up
The floor of the Caloris Basin – an impact crater created about 4 billion years ago – has risen since it was formed, suggesting that an active interior within Mercury pushed the floor up.

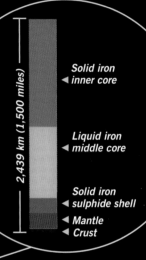

Solid iron inner core

Liquid iron middle core

Solid iron sulphide shell

Mantle
Crust

2,439 km (1,500 miles)

Solid iron inner core

Day side

Crust

430°C (800°F)

-160°C (-260°F)

A MOST UNUSUAL CORE

Gravitational measurements from *MESSENGER* suggest that Mercury's iron core is even bigger than previously thought. *MESSENGER* has also revealed the iron core has an unusually complex structure. The area above it is much denser than the rocky mantle, although not as dense as iron, which could mean that the central core is encased in a shell of iron sulphide 200 km (125 miles) thick – a situation not seen in any other planet.

ROCKY CORES AS A PERCENTAGE OF DIAMETER

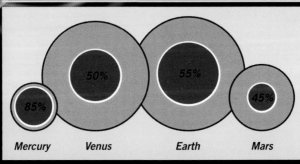

50%

55%

85%

45%

Mercury   Venus   Earth   Mars

Compared with the other rocky planets, Mercury has an unusually large core for its size. In fact, it would be fair to say that it is more core than planet. The core accounts for 42 per cent of the planet's volume – Earth's core is only 17 per cent of its volume.

HOW TO
CATCH A COMET

MAN HAS ALWAYS HUNTED.
Even before prehistory ditched the "pre" part of its name and became just history, hunters were using long and **sharp pointy things** to spear fish. But sometimes the fish slipped off the end. Then some bright spark had the idea of putting a barbed end on the sharp pointy thing and **the harpoon was born.** For centuries, the harpoon was the weapon of choice for hunting at sea but, lately, it has fallen out of vogue. NASA is planning to rehabilitate the harpoon but, instead of hunting whales at sea, **they will be hunting comets in space.**

Astronomers are fascinated by comets. These frozen chunks of dust and ice were formed when the Solar System was still a baby (that is well before history, prehistory, or any other sort of history) and they have remained unchanged ever since. As such, **they are like frozen time capsules**, crammed full of information about the origins of the Solar System.

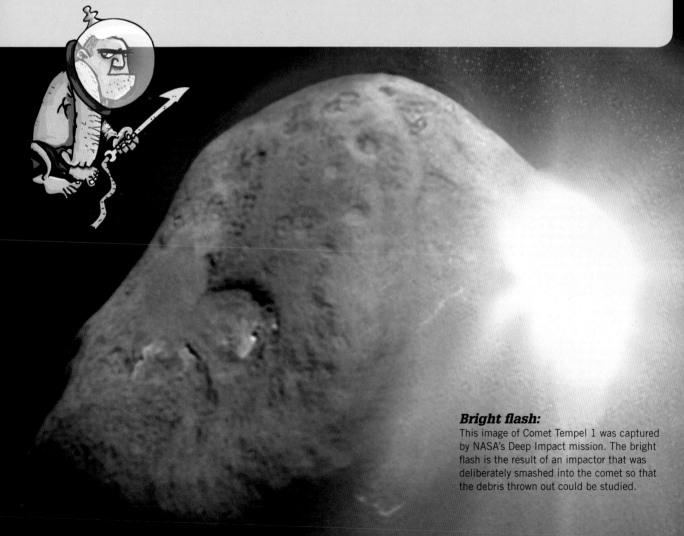

Bright flash:
This image of Comet Tempel 1 was captured by NASA's Deep Impact mission. The bright flash is the result of an impactor that was deliberately smashed into the comet so that the debris thrown out could be studied.

Astronomers would love to get **their hands on a sample of comet and unlock its secrets.** To make their wish a reality, NASA will be equipping a comet-hunting spacecraft called *OSIRIS-REx* with a harpoon and, to complete the historical synergy, they will fire it from a crossbow.

A comet can move through space quite quickly – about 240,000 kph (150,000 mph) – so landing a craft on its surface is a bit tricky. **The craft, which is slated to launch in 2016, will use a 2 m (6.5 ft) crossbow to fire a high-speed harpoon** with a special hollowed-out tip into the comet's surface.

The harpoon will grab a sample from inside the comet and then the sample will be winched back to *OSIRIS-REx* and returned to Earth. However, comet hunting is not as straightforward as you would think. So, "Call me Ishmael" and check out our comet-hunting guide.

THE HUNTER'S GUIDE TO COMETS...

Comets can be tricky little blighters but, like all "big game" hunts, knowing your quarry is half the battle. Firstly, you need to know where to find them. Like male lions, comets are kicked out of their homeland to wander alone. They come from two regions.

THE OORT CLOUD AND THE KUIPER BELT

The Oort Cloud is an immense spherical cloud of rocky debris left over from the formation of the Solar System. The Kuiper Belt is a disc-shaped region of icy debris just beyond Neptune's orbit.

The Oort Cloud extends about 30 trillion km (6.2 trillion miles) from the Sun.

Planetary region

The Kuiper Belt can be found about 6–7.5 billion km (3.7–4.6 billion miles) from the Sun.

Anatomy of a comet

1 Nucleus
Always aim for the heart. A comet's heart is a "dirty snowball" of water ice, frozen carbon dioxide, methane, and ammonia.

2 Jets of gas and dust
Gas and dust vent from the comet as the Sun vaporizes the comet's ice.

3 Coma
The coma is a cloud of dust and vapour that surrounds the nucleus.

4 Dust tail
A trail of dust particles.

5 Ion tail
The ion tail consists of charged particles pushed away from the comet by the solar wind, and can extend many tens of millions of kilometres from the comet.

EXPERT TIPS TO HELP YOU BAG YOUR PRIZE

Just because you know where to find them, don't think for one second that it's going to be plain sailing from here on out. Here's some dos and don'ts for any prospective hunter to keep in mind.

Spot the difference!

Dead ringer
Asteroids and comets are almost indistinguishable. Comets spend 99 per cent of their lives looking like asteroids.

Not very comet-y
Without the Sun's warmth, the water and gases that form the comet's tell-tale tail stay frozen solid – locked away in the comet's nucleus. In short, it doesn't look very comet-y.

This is a comet

This is an asteroid

DON'T BE FOOLED BY LOOKS
So you think you know a comet when you see one? Well think again. They only adopt their full comet-y plumage when they pass close to the Sun. Because comets have such huge orbits, they spend most of their time a long way from the Sun's warmth. So if you are not careful, a comet could pass right by you and you would never know.

Comet Holmes

Vented gas cloud
The cloud is created by gas that vents out of the comet when it is warmed by the Sun.

DO PREPARE YOURSELF
If you think you can wait for a comet to pass by before you shoot it, think again. Comets can come from deep space, which gives them plenty of time to pick up speed. You will either need to get in front of the comet to shoot it as it streaks towards you (not recommended), or you will need to match the comet's speed yourself.

DON'T BE TRICKED BY SIZE
When you do get your quarry in your sights, do not be fooled by its size – it is not as big as it looks. Most of what you can see is a cloud of gas, with the nucleus being a tiny speck somewhere in the middle. If you do not aim for the comet's tiny heart, your harpoon will just sail harmlessly through.

Comet Hale-Bopp

COMETS CAN MOVE 70 TIMES FASTER THAN A BULLET

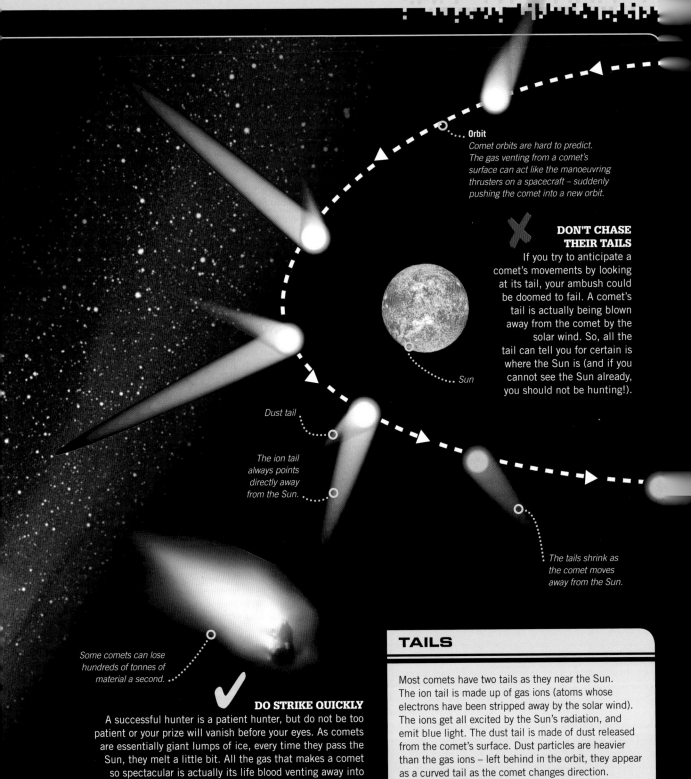

Orbit

Comet orbits are hard to predict. The gas venting from a comet's surface can act like the manoeuvring thrusters on a spacecraft – suddenly pushing the comet into a new orbit.

DON'T CHASE THEIR TAILS

If you try to anticipate a comet's movements by looking at its tail, your ambush could be doomed to fail. A comet's tail is actually being blown away from the comet by the solar wind. So, all the tail can tell you for certain is where the Sun is (and if you cannot see the Sun already, you should not be hunting!).

Sun

Dust tail

The ion tail always points directly away from the Sun.

The tails shrink as the comet moves away from the Sun.

Some comets can lose hundreds of tonnes of material a second.

DO STRIKE QUICKLY

A successful hunter is a patient hunter, but do not be too patient or your prize will vanish before your eyes. As comets are essentially giant lumps of ice, every time they pass the Sun, they melt a little bit. All the gas that makes a comet so spectacular is actually its life blood venting away into space. Eventually, all the ice and gas that holds it together will be gone and your comet will disintegrate.

TAILS

Most comets have two tails as they near the Sun. The ion tail is made up of gas ions (atoms whose electrons have been stripped away by the solar wind). The ions get all excited by the Sun's radiation, and emit blue light. The dust tail is made of dust released from the comet's surface. Dust particles are heavier than the gas ions – left behind in the orbit, they appear as a curved tail as the comet changes direction.

SATURN'S AMAZING RINGS

IN THE COURT OF THE PLANETS, red-eyed King Jupiter reigned supreme. Nothing rivalled his size, the violence of his atmosphere, the pull of his gravity, or the number of moons he held subject to his will. For billions of years his **only rival in the heavenly sphere was Saturn** who, although a gas giant himself, could never rival Jupiter's might.

So, like many a subordinate royal sibling, Saturn sought to outdo his relation in the only way he could. He gathered sparkling jewels, which he laid about himself in delicate rings; he became the dandy of the heavenly court; **he became the king of bling.**

But for millions of years his efforts went unnoticed by the peoples of the lowly rock-planet Earth, until one day, 400 years ago, a little Italian chap called **Galileo Galilei** turned a telescope to the heavens and proclaimed that Saturn had handles.

It took a few more years and some slightly more powerful telescopes before mankind identified Saturn's rings for what they were: **one of the most beautiful phenomena in the Solar System.** But still Saturn was not happy – he knew the stunning complexity of his finest decoration could never be appreciated from afar, so he brooded, awaiting the adulation his finery deserved.

Then, in 2004, he got his wish when a tiny machine sent by the humans to study his magnificence arrived. The machine was a probe called *Cassini* and it has revolutionized our understanding of Saturn and his glorious rings. It revealed the rings to be an elegantly complex system where **glittering beads of ice collide, reform, and collide again** (ensuring the rings stay nice and shiny).

It revealed **moons acting as shepherds,** keeping the rings in check. Other, tiny, moons pull material out into trailing tails in some places, and cut lines though the rings, while their gravity decorates the edges with waves in others. Meanwhile, vain Saturn's gravity **tears apart** any ice that clumps into pieces so large that they might threaten the aesthetic of his design.

F Ring
A dusty band of rubble orbiting 3,000 km (1,900 miles) beyond the main ring system. The nearby moon, Prometheus, distorts the ring by tens of kilometres as it passes.

A and B Rings
At 5–30 m (16–98 ft) deep, these are the densest parts of the ring system and are composed of 90–95 per cent water ice. One puzzle about the A and B Rings is that they are much redder than any of Saturn's moons. It is thought that tiny particles of iron oxide (rust) might be responsible.

Keeler Gap   Encke Gap

Cassini Division   Huygens Gap

F Ring   *A Ring*

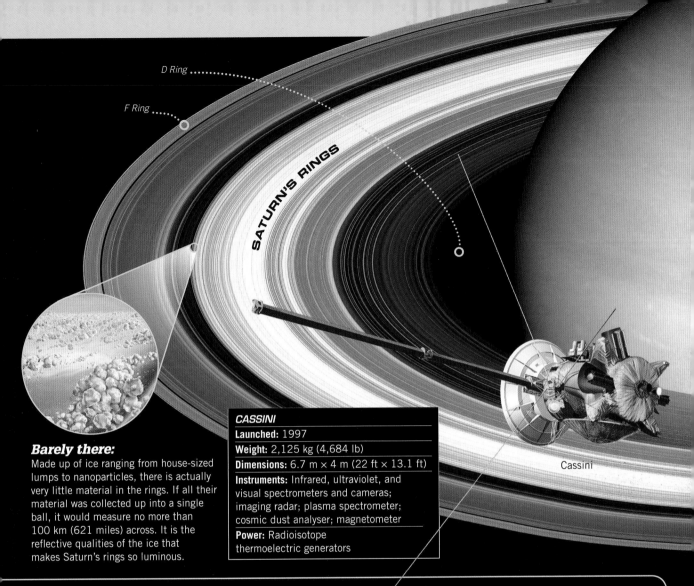

D Ring

F Ring

SATURN'S RINGS

Barely there:
Made up of ice ranging from house-sized lumps to nanoparticles, there is actually very little material in the rings. If all their material was collected up into a single ball, it would measure no more than 100 km (621 miles) across. It is the reflective qualities of the ice that makes Saturn's rings so luminous.

CASSINI	
Launched: 1997	
Weight: 2,125 kg (4,684 lb)	
Dimensions: 6.7 m × 4 m (22 ft × 13.1 ft)	
Instruments: Infrared, ultraviolet, and visual spectrometers and cameras; imaging radar; plasma spectrometer; cosmic dust analyser; magnetometer	
Power: Radioisotope thermoelectric generators	

Cassini

RINGS' ANATOMY

The main rings consist of thousands of ringlets, with each ringlet made up of millions of particles of ice, each an individual satellite orbiting the planet at up to 80,000 kph (50,000 mph). The rings are split by large gaps, which are caused by the gravitational influence of some of Saturn's moons orbiting outside the ring system.

C Ring
A wide, but faint ring that is only about 5 m (16 ft) deep. Its particles are known to be dirtier than the other rings, which is probably caused by pollution from meteoroids.

Saturn

D Ring
Saturn's innermost ring is its thinnest and faintest. It extends into Saturn's highest atmospheric clouds.

Maxwell Gap

Colombo Gap

B Ring

C Ring

Cross-section of Saturn's rings

92,000 km (57,166 miles)

Diameter of Earth: 12,756 km (7,926 miles)

74,000 km (46,000 miles; from Saturn's centre)

THE SEARCH
FOR ALIEN LIFE

IN THE 16TH CENTURY, the Italian philosopher-cum-astronomer, Giordano Bruno, speculated that stars (which, at that time, were thought to be little more than God's way of decorating the firmament) were in fact suns, around which might be **"... an infinity of worlds of the same kind as our own".**

Sadly, Bruno's heretical speculations led to him being **burned at the stake** by the Catholic Inquisition (not solely for his astronomical thinking) but he would not be the last great thinker to suggest that **Earth may not be the only world in the heavens capable of nurturing life** – English scientist Isaac Newton had similar thoughts 100 years later.

It was not until the early 1990s that the existence of the first extra-solar planet (exoplanet) was definitively confirmed – although unconfirmed discoveries had been staggering in since 1988. Since then, **confirmed exoplanet discoveries have risen into the thousands,** with thousands more awaiting confirmation.

The first such planets to be found were mostly super-sized Jupiter-like worlds with little hope of harbouring anything but the most robust single-celled alien life. But in recent years, improved detection techniques and a new generation of space telescopes are allowing scientists to build a catalogue of **Earth-like worlds that might just be capable of supporting alien life.** The most prolific of these exoplanet hunters is NASA's Kepler Space Telescope.

Since its launch in 2009, **Kepler has spotted thousands of exoplanet candidates** (planets awaiting independent confirmation). Just a few hundred of those are Earth-sized worlds, which are outnumbered by the so-called "super-Earth" planets (rocky planets with up to ten times the mass of Earth).

But, when it comes to searching for extra-terrestrials, size and composition are not everything – **if a planet lacks liquid water and a stable atmosphere, chances are it cannot support life.** A planet that orbits its parent star too closely will be too hot, and one whose orbit carries it too far away from its star will be too cold for water to exist in its life-supporting liquid state. We call the bit in between the **"Goldilocks zone",** and when we strip away all the discovered exoplanets that fall outside it, we are left with just a few confirmed potentially habitable worlds.

One reason for the ambiguity in the status of many of these discoveries is distance. Telescopes like Kepler search for exoplanets in orbit around stars that can be thousands of light-years away from Earth. At this distance, a star (a giant nuclear furnace whose surface is burning at tens of thousands of degrees) is quite

PSR B1257+12B:
The first confirmed discovery of an exoplanet was made in 1992. This world orbits a pulsar 1,000 light-years from Earth.

easy to spot but finding a planet (a small lump of rock) is a much more complicated prospect. To make matters worse, that tiny dark planet is being outshone by the star it orbits. In short, finding an Earth-like exoplanet is like trying to spot a mosquito as it flies across a floodlight several miles away. Kepler does this by looking for the almost imperceptible dimming that occurs when the planetary mosquito passes across its stellar floodlight. From the amount of dimming it detects, **astronomers can infer the existence of a planet and estimate its mass** – for a planet the size of Earth, that dimming might be as little as 0.004 per cent.

But **not all exoplanets are polite enough to pass in front of their parent star** (called a transit) as we look at it from Earth. For these tricky little blighters, more subtle techniques are required. Europe's dedicated planet-hunter, the High Accuracy Radial Velocity Planet Searcher (HARPS) – an instrument attached to an Earth-based telescope in Chile – **can detect the tiny "wobble" that occurs in a star's motion when it has a planet orbiting it.** Imagine a star and its orbiting planet to be two spinning skaters with unequal masses (with the star being the more massive one). If they link arms, the smaller one goes around in a larger circle, but can cause the more massive one to be thrown slightly out of balance as it spins. A planet orbiting a star is just like that added weight – its **gravitational pull causes the star to "wobble" slightly.** This wobble means that the star is moving back and forth relative to Earth. When the wobble carries the star away from us, the wavelength of its light is stretched towards the red end of the spectrum (redshift); when the star wobbles towards us, the light's wavelength is squeezed into the blue end of the spectrum (blueshift). The more massive the planet, the greater the shift – but, for small Earth-like exoplanets, the amount of shift is barely perceptible. Luckily,

HD 40307g:
This artist's impression of an Earth-like exoplanet may resemble HD 40307g, discovered in 2012, which could be the best candidate for alien life yet found.

THE OLDEST EXOPLANET DISCOVERED SO FAR, PSR B1620-26B, IS ALMOST 13 BILLION YEARS OLD

HARPS is sensitive enough to track these spectral changes. In its nine years of service, **HARPS has found 75 exoplanets.** Most have been uninhabitable super-Earths or gas giants, but in January 2012, the HARPS team announced the discovery of an exoplanet that sits comfortably within its star's habitable zone. The planet, HD 40307g, is the outermost of six worlds in orbit around a star about 42 light-years away called HD 40307. It orbits its star at a similar distance as Earth is from the Sun, enjoys similar levels of solar energy, and is likely to be a rocky world about seven times the mass of Earth. **There is also a good chance that it will have oceans of liquid water and a stable atmosphere.** Though it is 42 light-years away from us, HD 40307g is, relatively speaking, on our cosmic doorstep. This puts HD 40307g in range of future life-seeking exoplanet hunters.

HOW TO **FIND AN EXOPLANET**

Finding a tiny planet in orbit around a star so distant that, from Earth, it appears as little more than a white speck is challenging enough. Determining whether that planet could possibly harbour life would seem impossible. Luckily, astronomers have a few tricks up their sleeves.

FIND A GOOD SPOT
A star field without any "bright" stars is needed so they do not blind the instruments. Kepler, for example, is searching the Cygnus star field where it is monitoring 100,000 stars. Cygnus is ideal because it contains many stars that are similar to our Sun.

Search grid

Cygnus star field

"ROGUE PLANETS" ARE WORLDS THAT HAVE BEEN EJECTED FROM THEIR PLANETARY SYSTEM, AND ARE DOOMED TO WANDER ALONE IN THE DARK

FIND THE PLANET
There are two main ways to find a planet. The first, mentioned earlier, is the transit method – where we wait for the planet to cross the star, causing it to dim slightly. The second involves the technique of astrometric detection, which detects radial velocity (the wobbling of the star). Both determine the presence of planets by looking for subtle changes in the appearance of the parent star.

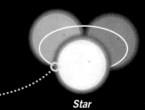

Radial velocity
As the star wobbles, it moves very slightly closer to, and further from, the orbiting exoplanet. By looking at the light's wavelength, we can tell if it is moving further away (as the wavelength will be stretched) or closer to us (as the wavelength will be compressed).

Star

Astrometric detection
As a planet orbits its star, the planet's gravity ever so slightly pulls the star from side to side – causing it to wobble.

Orbiting exoplanet

FIND OUT ITS MASS

Astronomers can use a technique called microlensing to work out the mass of a planet. Microlensing involves looking at a distant star and, when an exoplanet is in alignment with it, studying how the exoplanet's gravity affects the star's light.

1 Well, well, well

A planet (like all massive objects) creates a well in spacetime that we call gravity.

Apparent position of distant star

Light

Earth

Gravity well

Real position of distant star

2 Bending light

Einstein showed us that light is affected by gravity, just like anything else, and is bent as it passes a massive object.

Microlensing to work out a planet's mass

3 Shifting stars

By looking at distant stars, astronomers can figure out the planet's mass by how much it bends the stars' light – the more the stars seem to shift position, the greater the planet's mass.

LOCATION, LOCATION, LOCATION

Only exoplanets of the right size that lie within a distance of their parent star that permits the presence of liquid water are considered to be habitable. Kepler-22b is one such planet that is within its star's Goldilocks zone, and astronomers think that its environment may be similar to Earth's. This may not be enough, however, as the same could be said of Mars, which doesn't harbour life. A planet with an atmosphere that has evidence of water (H_2O), oxygen (O_2), and carbon dioxide (CO_2) could support life as we know it.

Mercury **Venus** **Earth** **Mars**

Solar System

Habitable zone

Kepler-22b System

Life, Jim, as we know it

The final confirmation of whether or not Kepler-22b is supporting life will be evidence of chemical biosignatures. Abundant life will affect the chemistry of the planet's atmosphere. For example, lots of vegetation will absorb light in certain recognizable wavelengths.

Kepler-22b

THE HOSTILE
BLUE PLANET

FLUNG AROUND its parent star at 400,000 kph (248,549 mph), the exoplanet HD 189733b is so close to the stellar furnace that its year lasts just 2.2 Earth days. Flayed by solar winds, **its atmosphere is stripped away and blasted into space by extreme ultraviolet and X-ray radiation** at the rate of 1,000 tonnes every second. The scorched atmosphere that survives the onslaught is rent by 7,000 kph (4,350 mph) winds laden with silicate particles, which become **supersonic shards of molten glass, propelled sideways.**

Indeed, if there is one exoplanet that deserves to have the blues, it is HD 189733b, which is rather apt as it has become the **first exoplanet to have its true colour determined** and, in line with that rather laboured set-up, that colour is blue.

Located about 63 light-years away, HD 189733b (which we will call **Howlin' Dave**) is (relatively speaking) right on our cosmic doorstep. It also happens to be one of the closest exoplanets that can be seen crossing the face of its star – making it the **most studied of all the alien planets.**

As a gas giant called a "hot Jupiter", Howlin' Dave is typical of most exoplanet discoveries – being big and hot (close to their stars) makes them relatively easy to spot. But this is **the first time astronomers have been able to measure an exoplanet's colour** and imagine how it would actually look through the window if you were able to fly past.

The planet's colour, which has been described as a **"deep cobalt blue",** is thought to come from clouds laden with reflective particles that contain silicon – raindrops of molten glass that scatter blue wavelengths of light. It adds to a growing portrait of Howlin' Dave.

In 2007, scientists using NASA's Spitzer Space Telescope studied Howlin' Dave and produced one of the first temperature maps of an exoplanet. It revealed a Janus-like world, with **one face tidally locked in a permanent furnace-facing gaze and the other hidden in eternal darkness.** The two sides differ in temperature by hundreds of degrees, driving the atmospheric turbulence that results in the planet's extreme winds.

In 2012, NASA's Swift satellite saw hydrogen atoms being torn from the planet's atmosphere at 482,803 kph (300,000 mph) by powerful solar winds – that is what happens when you orbit just 4 million km (2.4 million miles) from your star, Dave. Earth orbits the Sun at a far more sensible 150 million km (93 million miles).

In 2013, astronomers turned the European Southern Observatory's pragmatically named Very Large Telescope on the planet and detected **water molecules in its atmosphere – the first time good old H_2O had been found on an exoplanet**.

So Howlin' Dave is a bit like the stars of one of those "look at me, I'm a freak!" programmes – he was dealt a bad hand, but he is scientifically interesting and he is getting his five minutes of fame.

HD 189733b:
As the first exoplanet to have its true colour measured, HD 189733b's peaceful azure hue belies its true turbulent nature.

WE WANT ALIENS!

The golden egg of exoplanet research would be the discovery of a planet capable of supporting life. Recent estimates have increased the number of potential habitable-zone planets in the Milky Way to as many as 100 billion. Of course, there is a difference to being "in the habitable zone" and actually being habitable. Venus sits neatly within the Sun's habitable zone, but runaway global warming (caused by a toxic carbon-dioxide-rich atmosphere 93 times denser than Earth's) has left the planet with a surface that can reach a toasty 460°C (860°F).

THE FIERCE WINDS ON HD 189733B CAN REACH TEMPERATURES OF 1,000°C (1,800°F)

HOW FAR AWAY?

At a mere 63 light-years away, HD 189733b is close to us in astronomical terms, but that is still quite a long way. One light-year is 9.46 trillion km (5.88 trillion miles), which means that Howlin' Dave is 596 trillion km (370 trillion miles) away. To get an idea of the distance, if we scaled down the galaxy so that the Sun was reduced to the size of a tennis ball, Earth (shrunk to the size of a grain of sand) would be 8 m (26 ft) away. HD 189733b (shrunk to the size of a cherry stone) would be 31,870 km (19,803 miles) away – almost six times the distance from London to New York.

The area of sky covered in this image is roughly equivalent to the width of your little finger held at arm's length.

This artist's impression shows HD 189733b in orbit around its star.

Hubble's view: Shown here is Howlin' Dave's parent star, HD 189733, as seen by Hubble. That we know anything about a planet that orbits this tiny white speck is impressive – that we know so much is pretty awesome.

THE PLANET HUNTER

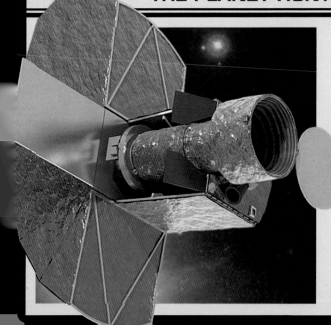

The European Space Agency (ESA) has just authorized the development of a new planet-hunting satellite mission. Due for launch in 2017, the CHaracterising ExOPlanet Satellite (CHEOPS) will be tasked with building a catalogue of potential exoplanets for future life-detecting missions to target. Scientists have estimated that, in our galaxy alone, there are tens of billions of rocky, Earth-sized planets, many of which are lying inside their stars' Goldilocks zone – meaning the existence of life beyond Earth is not just possible, but that it might be inevitable and common. Whether any of that life is anything other than self-replicating slime is another matter but, given the almost incomprehensible scale of the Universe, the odds of there being a "proper ET" somewhere out there look pretty healthy.

CHEOPS: Due to begin operations in 2017, ESA's CHEOPS planet-hunter will build a catalogue of potentially life-supporting worlds for future life-hunters to exploit.

COLOUR ANALYSIS

Every chemical element absorbs and emits certain frequencies of light. Splitting starlight into a spectrum reveals lines that correspond to the elements of which the star is made – like a chemical barcode. The same technique, called spectroscopy, can also be used, with appropriate cunning, to work out the composition of the atmosphere of an exoplanet, and sometimes even something about its surface and its true colour.

THE NEXT GENERATION OF TELESCOPES WILL ALLOW ASTRONOMERS TO SEARCH EXOPLANET ATMOSPHERES FOR THE CHEMICAL SIGNATURES OF LIFE

HOW TO STUDY AN EXOPLANET

Until we have telescopes powerful enough to capture exoplanets directly, the best way to learn anything about them is to see what happens when one makes a transit of its parent star. The light obtained from the planet is then used to figure out what elements are in its atmosphere, and also what its true colour is.

True colour
When Howlin' Dave had passed behind his star, the overall light spectrum obtained was analysed and seen to drop in the blue part of the spectrum, meaning Dave is blue.

HD 189733b

HD 189733

Spectral lines
The complete spectrum for the planet will have many spectral lines. By looking for each element's specific barcode within the spectrum (in this case oxygen), astronomers can identify the elements present.

Blocked light
The amount of light the planet blocks during a transit can be as low as 0.004% (Howlin' Dave blocks about 2%).

THE SPACE ROCK
THAT "KILLED" PLUTO

BACK IN 2006, the world of astronomy was **torn asunder** by the controversial decision to **demote Pluto.** Overnight, the planet was stripped of its status, and became a dwarf planet. Many astronomers, who felt the decision had been usurped by a minority, were furious (as were countless students who laboured to memorize the planetary mnemonic "My Very Educated Mother Just Served Us Nine Pizzas").

The trouble started back in 2005, when astronomers discovered another world hiding in the darkness of the Kuiper Belt that seemed to be **even bigger than Pluto.** The world was named "Eris" – after the Greek goddess of chaos and strife – and it has been living up to its name ever since.

Its discovery sparked a debate about the definition of a planet that could have seen the number of "planets" in our Solar System swell to 14 or more, but instead, in 2006, it saw Pluto being chopped off the end of the planetary roll call. **The arguments have been raging ever since.**

Then, in 2010, astronomers measured Eris's girth with greater accuracy and, yet again, it caused trouble. Eris, as it turns out, is actually **significantly smaller than was first estimated** – small enough to pass the Kuiper Belt crown back to Pluto.

The difference is tiny: Eris is a mere 4 km (2.5 miles) narrower than Pluto. But it was enough for the "promote Pluto" trumpets to start sounding once again. Unfortunately, the planet's demotion was not based on its Kuiper-ranking alone, and all the other reasons for its fall from grace still stand – **its wacky orbit** and its tiny size relative to "real" planets.

Also, given the margin of error that comes with measuring a tiny, dark object 39 times further from the Sun than Earth, its size-ranking could still change. Intriguingly, the new measurements have shown Eris to be 25 per cent more massive (the term "weight" does not apply in space) than Pluto – **implying that Eris contains more rock than icy Pluto.**

So, until something changes, the new mnemonic in Pluto's absence is "Mean Very Evil Men Just Shortened Up Nature".

THE DWARF PLANET SEDNA IS SO FAR AWAY THAT IT TAKES 11,400 YEARS TO MAKE ONE ORBIT OF THE SUN

Eris:
The architect of Pluto's downfall, Eris, looks back on the distant Sun in this artist's impression.

NEW HORIZONS

After nearly ten years and 5 billion km (3 billion miles), NASA's *New Horizons* spacecraft is due to fly past Pluto and its moons in July 2015. It will provide us with our first close-up study of Pluto and will, finally, determine its size once and for all. It will then continue on into the Kuiper Belt.

Launched: 1997

Weight: 465 kg (1,025 lb)

Width: 2.5 m (8.2 ft)

Power: Nuclear (radioisotope thermoelectric generator)

Max speed: 72,420 kph (45,000 mph)

MEET THE DWARF PLANETS

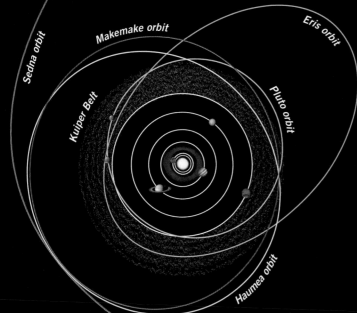

Sedna orbit

Makemake orbit

Eris orbit

Kuiper Belt

Pluto orbit

Haumea orbit

1. PLUTO
Discovered: 1930
Diameter: 2,344 km (1,457 miles)

Pluto is named after the ancient Roman god who ruled the underworld.

2. ERIS
Discovered: 2005
Diameter: 2,340 km (1,454 miles)

Eris is the ancient Greek goddess of chaos, strife, and discord.

3. HAUMEA
Discovered: 2003
Diameter: 1,960 km (1,218 miles; at its widest point)

This egg-shaped rock is named after a Hawaiian fertility god.

4. MAKEMEA
Discovered: 2005
Diameter: approx. 1,900 km (1,180 miles)

The Polynesian god of fertility gives this rock its name.

5. SEDNA
Discovered: 2004
Diameter: 1,180–1,800 km (733–1,118 miles)

The Inuit goddess of the sea gives this dwarf planet its name.

The Kuiper Belt is often called our Solar System's final frontier. It is a disk-shaped region of icy debris about 6–7.5 billion km (3.7–4.6 million miles) from the Sun. Over 1,000 Kuiper Belt objects have been discovered in the Belt, almost all of them since 1992.

Sedna's orbit makes it one of the most distant objects in the Solar System

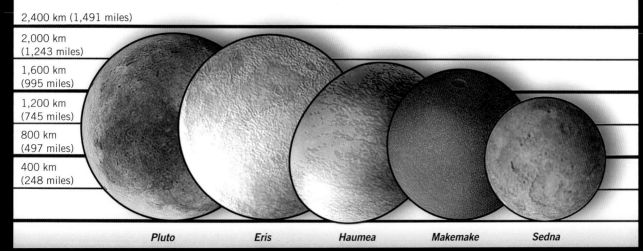

2,400 km (1,491 miles)
2,000 km (1,243 miles)
1,600 km (995 miles)
1,200 km (745 miles)
800 km (497 miles)
400 km (248 miles)

Pluto   Eris   Haumea   Makemake   Sedna

The unusual suspects: Thousands of rocks, one line-up, no planets.

MEASURING SOMETHING
YOU CAN BARELY SEE

In the past few years, new data seems to have revealed that Eris might actually be smaller than Pluto after all. But is it really? And how do we know? Let's investigate!

HOW HAS ERIS SHRUNK?

It has not: its size was just overestimated. The first measurements of Eris's size were based on how bright it appeared and the amount of light it reflects. This gave a diameter of about 2,400 km (1,490 miles) – bigger than Pluto.

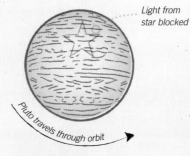

Distant star

Pluto

Light from star blocked

Pluto travels through orbit

Star visible again

1 ***Distant star***
Astronomers look for a distant star on the far side of Pluto.

2 ***Blocking light***
As Pluto moves through its orbit it crosses the star, blocking its light.

3 ***Diameter measured***
By measuring how long it takes for the star to reappear, astronomers can calculate Pluto's diameter.

SO WHAT HAPPENED?

Recently astronomers had a chance to measure a stellar occultation (above), which is much more accurate. Astronomers need to measure at least two instances of stellar occultation, from different locations, to accurately gauge an object's size. Using this method, the new measurement for Eris is a far smaller diameter of about 2,340 km (1,454 miles).

SO ERIS IS DEFINITELY SMALLER THAN PLUTO?

Erm... no. Even though the generally accepted measurement for Pluto is 2,400 km (1,490 miles), this is far from definitive. The trouble with using the occultation method is that Pluto has a light atmosphere, which is enough to mess around

with occultation measurements. Pluto's atmosphere therefore makes the planet appear bigger than it is (since an occultation measurement relies on the amount of light being blocked), so the information we have is probably misleading. Dang!

Atmospheric methane
Sunlight causes the methane in Pluto's atmosphere to decompose into opaque hydrocarbons.

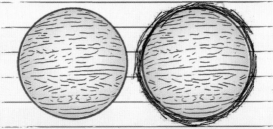

Without opaque atmosphere

With opaque atmosphere

THE FIRST HUMAN IN SPACE

PIONEER 10.
THE LITTLE SPACECRAFT
THAT COULD

ENGAGE WARP DRIVE!

GRAVITY LENSING
TO SEE THE COSMOS

MAPPING THE MILKY WAY

IS THERE LIFE ON MARS?

COLONIZING MARS

SPACE: THE FATAL FRONTIER

ESA'S *ROSETTA* COMET CHASER

VOYAGER: OUR DISTANT EMISSARY

TO BOLDLY GO

LOOKING BEYOND MARS FOR LIFE

DETECTING KILLER ASTEROIDS

A WEBB TO CATCH THE OLDEST STARS

THE FIRST
HUMAN IN SPACE

AT 6:07 AM ON 12 APRIL 1961, Yuri Gagarin, a 26-year-old astronaut from the USSR, left the surface of Earth and travelled into space – **becoming the first human to escape the confines of our planet and extend humankind's reach into the heavens.**

The announcement left the US space programme reeling. The USSR's **"Space Race"** rival was due to launch its first human into space a few weeks later, but their astronaut, Alan Shepard, had to settle for being the first American in space.

Cosmonaut Yuri Gagarin in the cockpit of his **Vostok 1** *spaceship*

"CIRCLING THE EARTH IN MY ORBITAL SPACESHIP, I MARVELLED AT THE BEAUTY OF OUR PLANET... LET US SAFEGUARD AND ENHANCE THIS BEAUTY, NOT DESTROY IT"

YURI GAGARIN

Vostok-K: Gagarin's *Vostok 1* craft was launched into orbit on a Vostok-K rocket. The Vostok-K had failed on its first launch just four months before Gagarin's mission.

The launch into the unknown from Baikonur, Kazakhstan

Gagarin's five-tonne spaceship, *Vostok 1,* was **carried into space on board a converted ballistic missile,** which propelled the plucky space-navigator to speeds in excess of 27,000 kph (17,000 mph) from launch in a remote region of Kazakhstan.

At the heady altitude of 327 km (203 miles), his craft then proceeded to orbit Earth – a journey that took just 89 minutes to complete. **A mere 108 minutes after leaving the planet in obscurity, Gagarin returned safely to Earth as a national hero** and an international celebrity.

Unknown to Gagarin, during launch, the second-stage of the rocket burned for longer than planned – **thrusting the *Vostok 1* orbiter into a higher orbit than was intended.** This meant that, had his braking engine failed, it would have taken Gagarin's craft 15 days to fall back to Earth, as there was no back-up. This would have been five days longer than his food and life-support system would have allowed.

Nor was the return to Earth as smooth as intended. During re-entry, a valve within the braking engine failed to close completely, which let some fuel escape –

causing the engine to shut down a second too early. Gases were vented that caused the craft to enter into a violent spin. Also, the technical module failed to separate completely from the re-entry section.

Fortunately, the spin subsided and the heat created during re-entry burned through the cable that still connected the technical module – **allowing Gagarin to jettison the craft's main hatch and eject from the vehicle** at an altitude of 7 km (4.3 miles), and return safely to terra firma.

A VOYAGE AROUND THE WORLD...

Yuri Gagarin was selected from an elite group of Soviet pilots, known as the "Sochi Six", to become the first human to be launched into space and orbit Earth. Despite his piloting pedigree, Gagarin was really just a passenger on board the *Vostok 1* spacecraft – because of the uncertainty about how space flight would affect him, the craft was controlled remotely from Earth. His flight lasted just 108 minutes but, in that time, he orbited the planet, saw the sun rise and set and, most importantly, landed back on Earth alive and well.

LAIKA

In November 1957, a stray dog called Laika became the first living thing to be sent into space. Sent by the USSR, the plucky hound safely orbited Earth for seven days, proving that it was possible for a creature to survive a launch and live in space. Unfortunately, there was no plan to bring her back and, after seven days, Laika was put to sleep before her oxygen supply ran out.

VOSTOK 1

The craft was made up of two modules – a 2.3 m (7.5 ft) diameter re-entry capsule and an equipment module. In the cabin, there was an envelope that contained a special code, which the cosmonaut could use to override the automated computer system in case of emergency. The spherical re-entry capsule was weighted so it would roll into position to ensure the craft was pointing the right way during re-entry.

Whip antenna ....

Communications antenna

Retro engine (used for braking)

SPACE RACE

On 4 October 1957, the world (and USA in particular) was stunned by the news that USSR had successfully launched the world's first orbiting satellite, *Sputnik 1*, into space. The Space Race had begun. In the early years, the race was the USSR's to lose – after *Sputnik*, it put the first living creature into Earth orbit (Laika) and then the first human (Gagarin). The USA, determined it would achieve the ultimate first, threw everything at the race to put the first man on the Moon...

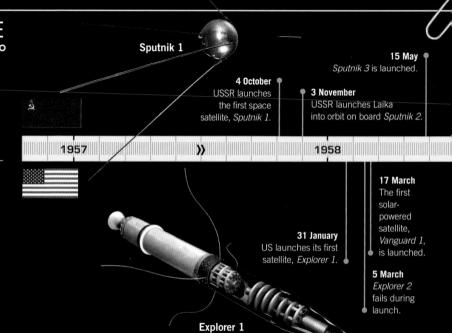

Sputnik 1

4 October
USSR launches the first space satellite, *Sputnik 1*.

3 November
USSR launches Laika into orbit on board *Sputnik 2*.

15 May
Sputnik 3 is launched.

1957   »   1958

31 January
US launches its first satellite, *Explorer 1*.

17 March
The first solar-powered satellite, *Vanguard 1*, is launched.

5 March
Explorer 2 fails during launch.

Explorer 1

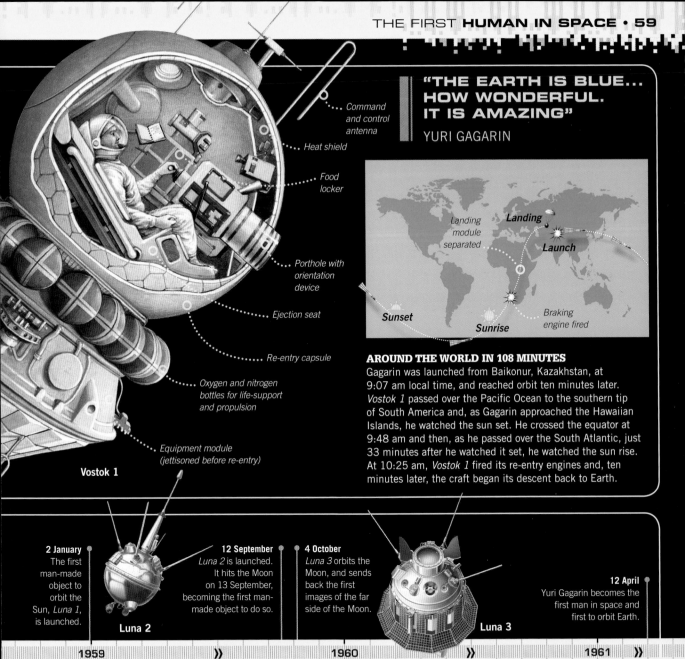

Vostok 1

Command and control antenna

Heat shield

Food locker

Porthole with orientation device

Ejection seat

Re-entry capsule

Oxygen and nitrogen bottles for life-support and propulsion

Equipment module (jettisoned before re-entry)

"THE EARTH IS BLUE... HOW WONDERFUL. IT IS AMAZING"

YURI GAGARIN

Landing module separated

Landing

Launch

Sunset

Sunrise

Braking engine fired

AROUND THE WORLD IN 108 MINUTES

Gagarin was launched from Baikonur, Kazakhstan, at 9:07 am local time, and reached orbit ten minutes later. *Vostok 1* passed over the Pacific Ocean to the southern tip of South America and, as Gagarin approached the Hawaiian Islands, he watched the sun set. He crossed the equator at 9:48 am and then, as he passed over the South Atlantic, just 33 minutes after he watched it set, he watched the sun rise. At 10:25 am, *Vostok 1* fired its re-entry engines and, ten minutes later, the craft began its descent back to Earth.

2 January
The first man-made object to orbit the Sun, *Luna 1*, is launched.

Luna 2

12 September
Luna 2 is launched. It hits the Moon on 13 September, becoming the first man-made object to do so.

4 October
Luna 3 orbits the Moon, and sends back the first images of the far side of the Moon.

Luna 3

12 April
Yuri Gagarin becomes the first man in space and first to orbit Earth.

1959   »   1960   »   1961   »

11 October
NASA's first spacecraft, *Pioneer 1*, is launched.

1 October
The National Aeronautics and Space Administration (NASA) is formed.

2 April
NASA selects its first group of astronauts, dubbed the "Mercury Seven".

3 March
Pioneer 4 passes within 60,000 km (37,500 miles) of the Moon's surface.

Pioneer 4

1 April
The first successful weather satellite, *Tiros 1*, is launched.

18 August
The first camera-equipped spy satellite, *Discoverer XIV*, is launched.

8 November
John F Kennedy is elected as the 35th President of the United States, and sets NASA the goal of landing a man on the Moon before 1970, a goal achieved in 1969.

PIONEER 10:
THE LITTLE SPACECRAFT THAT COULD

Helium vector magnetometer
Measures and maps Jupiter's magnetic field.

UNTIL ABOUT 40 YEARS AGO, the furthest any man-made object had ventured into space was Mars, but it was clear that **going further would present a series of challenges.** Beyond Mars, there lay an 180 million km (110 million mile) wide barrier of rocks of all sizes, barrelling through space at tens of thousands of kilometres per hour – called the asteroid belt. Any craft that ventured in could be damaged by huge rocks, or possibly be **pelted with tiny rocks that could wreck its instruments.**

Then, a little over 42 years ago, NASA put the theory to the test. Launched on 2 March 1972, *Pioneer 10* left Earth on a mission to study Jupiter. To reach the planet, it would have to **traverse the asteroid belt.** A few months later, *Pioneer 10* entered the belt but, instead of being smashed to a metallic pulp, it sailed through without a hitch. It turned out that, far from being a densely packed highway of rocky death, the asteroid belt was mostly empty space. The Solar System was now ours to explore.

Pioneer 10 went on to become the **first man-made object to study Jupiter and the first to cross the orbits of Saturn, Neptune, Uranus, and Pluto.** Long after its intended 21-month lifespan had been exceeded, *Pioneer 10* kept on trucking until 2003, when, at the outer limits of our Solar System and 12.2 billion km (7.5 billion miles) from home, it sent its last transmission. *Pioneer 10* (and its sister craft *Pioneer 11*, launched in 1973 to visit Saturn) was one of the **great space adventures** and it paved the way for many more.

BUSTING OUT!

Until *Pioneer 10*, the furthest mankind had extended his reach was Mars. The *Pioneer* probes went much, much further...

IT WILL TAKE MORE THAN 2 MILLION YEARS FOR *PIONEER 10* TO PASS ALDEBARAN, THE NEAREST STAR ON ITS TRAJECTORY

Asteroid belt

Sun Earth | Jupiter | Saturn | Uranus | Sunlight takes 4 hours to reach here | Neptune

Distance from the Sun in astronomical units (AU – 1 AU = distance from Earth to the Sun) | *10 AU (1,500 million km/ 932 million miles)* | *20 AU (3 billion km/ 1.8 billion miles)* | *25 AU* | *30 AU*

Geiger tube telescope
*Studies the properties of
electrons and protons in
Jupiter's radiation belts* ·············

Trapped radiation detector
*Captures radiation and
analyses its properties*

Ultraviolet photometer
*Senses ultraviolet light
to determine how much
hydrogen and helium is
present on Jupiter and
in space.*

PIONEER

Launched
Pioneer 10: March 1972
Pioneer 11: April 1973
Mass (both craft): 259 kg (571 lb)
Antenna (both craft): 2.75 m (9 ft)

Plasma analyser
*Detects particles in
the solar wind*

**Asteroid/
meteoroid
detector**

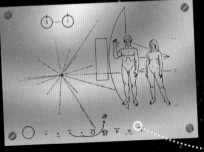

The Pioneer plaque:
Gold-anodized aluminium plates (above)
were fixed to both *Pioneer 10* and *Pioneer
11*. They were designed to communicate
the location and appearance of the human
race as well as information about the
spacecrafts' origins – in case the probes
were intercepted by extraterrestrials.

**Cosmic ray
telescope**

*This shows Earth's
position in the
Solar System, and
Pioneer's route.*

Infrared radiometer
*Provides information on
Jupiter's heat output*

Radioisotope thermoelectric generators
*These two generators use the
radioactive decay of Plutonium-238
to provide the craft with power.*

**Kuiper Belt
(full length of the
lighter green bar)**

Pluto   Haumea*   Makemake*

35 AU   40 AU (6 billion km/   45 AU   50 AU   55 AU   60 AU   65 AU
                3.5 billion miles)

*= Approximate distances

PIONEER'S
FANTASTIC VOYAGE

The voyage of the *Pioneer* probes was a truly epic achievement that revolutionized our understanding of the Solar System and paved the way for future robotic explorers, such as NASA's iconic *Voyager* missions.

1 ### Launch from Earth
Pioneer 10 was launched on a three-stage Atlas-Centaur rocket from Florida, USA, in March 1972. The craft reached 52,140 kph (32,400 mph), making it the fastest man-made object to leave Earth. At this speed, *Pioneer* could pass the Moon in 11 hours and cross the orbit of Mars in just 12 weeks.

2 ### Asteroid belt
Pioneer 10 entered the asteroid belt in July 1972. At the time, it was thought that the belt was densely populated with asteroids cannoning through space at 72,400 kph (45,000 mph). Scientists were worried that *Pioneer* would be unable to navigate safely through and, like a blind hedgehog on a motorway, would be smashed to smithereens.

3 ### Leaving the asteroid belt
Pioneer passed safely out of the asteroid belt in February 1973. It had shown the belt to be actually quite sparsely populated. The revelation opened the door to future deep-space exploration.

4 ### Jupiter
In December 1973, *Pioneer 10* passed by Jupiter, becoming the first craft to photograph and make direct observations of the red-eyed gas giant. *Pioneer 10* charted Jupiter's intense radiation belts, studied its magnetic field, and confirmed the fact that Jupiter radiated more heat than it absorbed from the Sun.

Pioneer 11
Launched in April 1973, Pioneer 11 studied the asteroid belt, Jupiter, and Saturn.

Saturn

Asteroid belt

Mercury

Uranus

Pioneer 10's image of Jupiter:
At its closest approach, the craft passed within 132,252 km (82,177 miles) of the planet's outer atmosphere. Under the pull of Jupiter's gravity, *Pioneer 10* was accelerated to 132,000 kph (82,000 mph).

Sunlight takes 10 hours to reach here

Sunlight takes 12 hours to reach here

Pioneer 11

Eris*

70 AU (10.5 billion km/ 6.5 billion miles)   75 AU   80 AU   85 AU   90 AU   95 AU   100

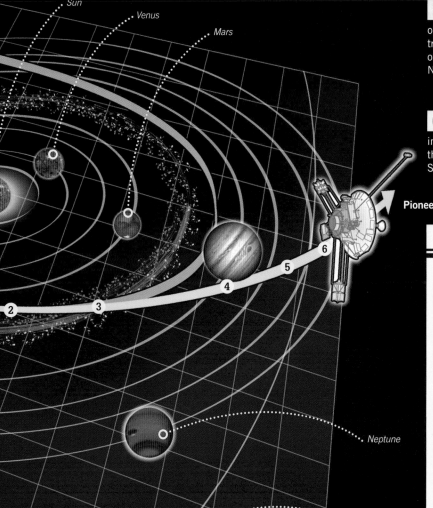

Sun
Venus
Mars

5 ▷ *Pluto*
Pioneer 10 became the first man-made object to pass the orbit of Pluto when it travelled past in April 1983. Pluto's irregular orbit meant it was closer to the Sun than Neptune in 1983.

6 ▷ *Neptune*
The craft crossed the orbit of Neptune in June 1983. Soon after, *Pioneer 10* became the first man-made object to depart the inner Solar System.

Pioneer 10

Neptune
Pluto

MISSION FACTS

• *Pioneer 10* continued to take readings of the outer regions of the Solar System until its science mission officially ended on 31 March 1997.
• On 27 April 2002, *Pioneer 10* sent its last decipherable signal.
• By April 2015, based on its last-known speed, *Pioneer 10* reached 112 AU, and *Pioneer 11* reached 92 AU.
• Each year, both *Pioneer* crafts travel about 5,000 km (3,100 miles) less than scientists calculate that they should. With no air to slow the crafts down (space, after all, is a vacuum), scientists have struggled to come up with an explanation for this anomaly. Proposed solutions have varied from the mundane – gas leakage from the craft or heat radiation – to the much more dramatic suggestion that this reveals flaws in our understanding of gravitational physics.

Pioneer 10

Sunlight takes
19 hours to reach here

105 AU   110 AU   115 AU   120 AU   125 AU   130 AU

= Approximate distances

VOYAGER: OUR DISTANT EMISSARY

IN THE LATE 1970S, an extremely rare event took place: the orbits of the outer planets of the Solar System – Jupiter, Saturn, Uranus, and Pluto – aligned in such a way that it would be possible for a pair of spacecraft to visit and study them. To take advantage of this **once-in-every-175** years opportunity, NASA launched the twin *Voyager* probes on a "grand tour of the planets" in 1977. No one could have guessed that, almost four decades later, the (by then) rickety old probes would **still be travelling and still be making discoveries** and pushing forward the boundaries of science.

VOYAGER

Launched
Voyager 1: 5 September 1977
Voyager 2: 20 August 1977

Mass: 721.9 kg (1,600 lb)

Current speed
Voyager 1: 62,000 kph (38,000 mph)
Voyager 2: 55,500 kph (34,500 mph)

Distance from Earth in 2015
Voyager 1: 19.3 billion km
(12 billion miles)
Voyager 2: 15.9 billion km
(9.9 billion miles)

Cameras and spectrometer

Cosmic ray detector

Artist's impression of Voyager probe

Star trackers

Magnetometer boom

High-gain antenna

TRAVELLING FAR

It may not seem it from all those artist's impressions of closely packed planets, but the Solar System is a vast place and, since 1977, the *Voyager* probes have travelled a long, long way...

Voyager *probes:*
The identical *Voyager 1* and *Voyager 2* probes were launched in 1977 to take advantage of a favourable alignment of the planets. They were designated to study the planetary systems of Jupiter and Saturn but, four decades later, they are still travelling through space.

| Sun | Earth | Asteroid belt | Jupiter | Saturn | Uranus | | Neptune |

Sunlight takes 4 hours to reach here

Distance from the Sun in astronomical units (AU, 1 AU = distance from Earth to the Sun)

10 AU (1,500 million km/ 932 million miles)

20 AU (3 billion km/ 1.8 billion miles)

25 AU

30 AU

Now, nearly four decades and about 19.3 billion km (12 billion miles) later, *Voyager 1* is leaving our Solar System behind and **passing into the dark, unexplored expanse of interstellar space.**

For some years now, data beamed back from *Voyager 1* data that takes more than 16 hours to reach Earth) has hinted that the venerable machine **might finally be passing the outer limits of our Solar System.** But things are not quite what scientists expected them to be.

into the interstellar medium like a bucket of water thrown against a wall.

In 2010, *Voyager 1* seemed to reach this point, but the craft's instruments indicated that the wind had just stopped dead – instead of a maelstrom of clashing solar particles, there was **just a stagnant pool of stationary particles.**

This countered everything that existing models of the Solar System had predicted. This led scientists to reassess how they think about the

particle accelerator that picks up particles from within the heliopause and whips them up into a high-energy frenzy.

Around 2012, the craft began to detect a dramatic drop in the number of solar particles it was finding and a huge increase in the amount of cosmic radiation, which suggested that the craft was about to become the first man-made object to leave the Solar System – **as it successfully did in 2013!**

A SIGNAL, TRAVELLING AT THE SPEED OF LIGHT, TAKES ABOUT 13 HOURS ONE WAY TO REACH *VOYAGER 2*, AND 16 HOURS TO REACH *VOYAGER 1*

Scientists define the limits of our Solar System as being the point at which the solar wind (a stream of charged particles flowing out from the Sun at supersonic speeds) runs out of puff – in other words, **the Solar System ends where the Sun's influence ends.** While it has the strength, the solar wind pushes against the gas and dust of interstellar space and inflates a giant "bubble" of charged particles and magnetic fields called the heliosphere. At the edge of this bubble, scientists had expected to find a pressure boundary called the heliopause, where the solar wind smashed

heliosphere, and **researchers suggested that *Voyager 1* was not as close to the interstellar boundary as suspected.**

Analysis of more data collected in 2010 found further anomalies at the edge of the heliosphere. Scientists had expected that, as solar wind slowed, **the heliosphere's magnetic field would fluctuate** and scramble any high-energy cosmic rays trying to pass through it. But as the magnetic field became more chaotic, the number of high-energy particles actually increased. Researchers then suggested that the magnetic field may actually be acting as a sort of

VOYAGER DISCS

Both of the *Voyager* crafts carry identical 12-inch gold-plated discs which include:
• 117 pictures of Earth, the Solar System, and various plants and animals.
• Greetings in 54 languages – including one in Mandarin, saying, "Hope everyone's well. We are thinking about you all. Please come here to visit when you have time" – and a brief hello from some humpback whales.

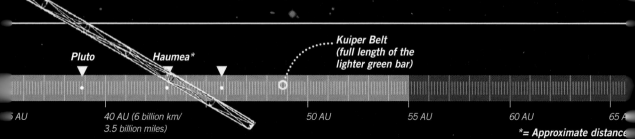

Pluto

Haumea*

Kuiper Belt
(full length of the
lighter green bar)

5 AU   40 AU (6 billion km/   50 AU   55 AU   60 AU   65 A
                3.5 billion miles)

*= Approximate distance

THE VOYAGE FROM HOME...

A rare planetary alignment made the *Voyager* probes' "grand tour" possible, but no one could have imagined how far they would travel.

1 Launch from Earth
Voyager 2 is launched in August 1977 – with *Voyager 1* following a few weeks later in September.

Moon

Earth

2 Say cheese!
Voyager 1 takes the first spacecraft photograph of the Moon and Earth in a single frame in September 1977.

3 Jupiter
Voyager 1 makes its closest approach to Jupiter in March 1979, followed by *Voyager 2* in July.

4 Saturn
Voyager 1 flies by Saturn in November 1980, with *Voyager 2* chugging past in August the following year.

5 Uranus
Voyager 2 becomes the first spacecraft to visit Uranus in January 1986.

6 Neptune
Voyager 2 becomes the first spacecraft to visit Neptune in August 1989.

Mercury

Solar System:
This graphic is heavily stylized and is not even slightly to scale. To get a better idea of the distances involved, have a look at the strip below (some of the key events have been marked in yellow).

Solar System

3

Jupiter:
This image of Jupiter's Great Red Spot was taken by *Voyager 1* in 1979 when the spacecraft was 9.2 million km (5.7 million miles) from the gas giant.

Sunlight takes 10 hours to reach here

Sunlight takes 12 hours to reach here

Termination shock

Eris*

8

70 AU (10.5 billion km/ 6.5 billion miles)   75 AU   80 AU   85 AU   90 AU   95 AU   10

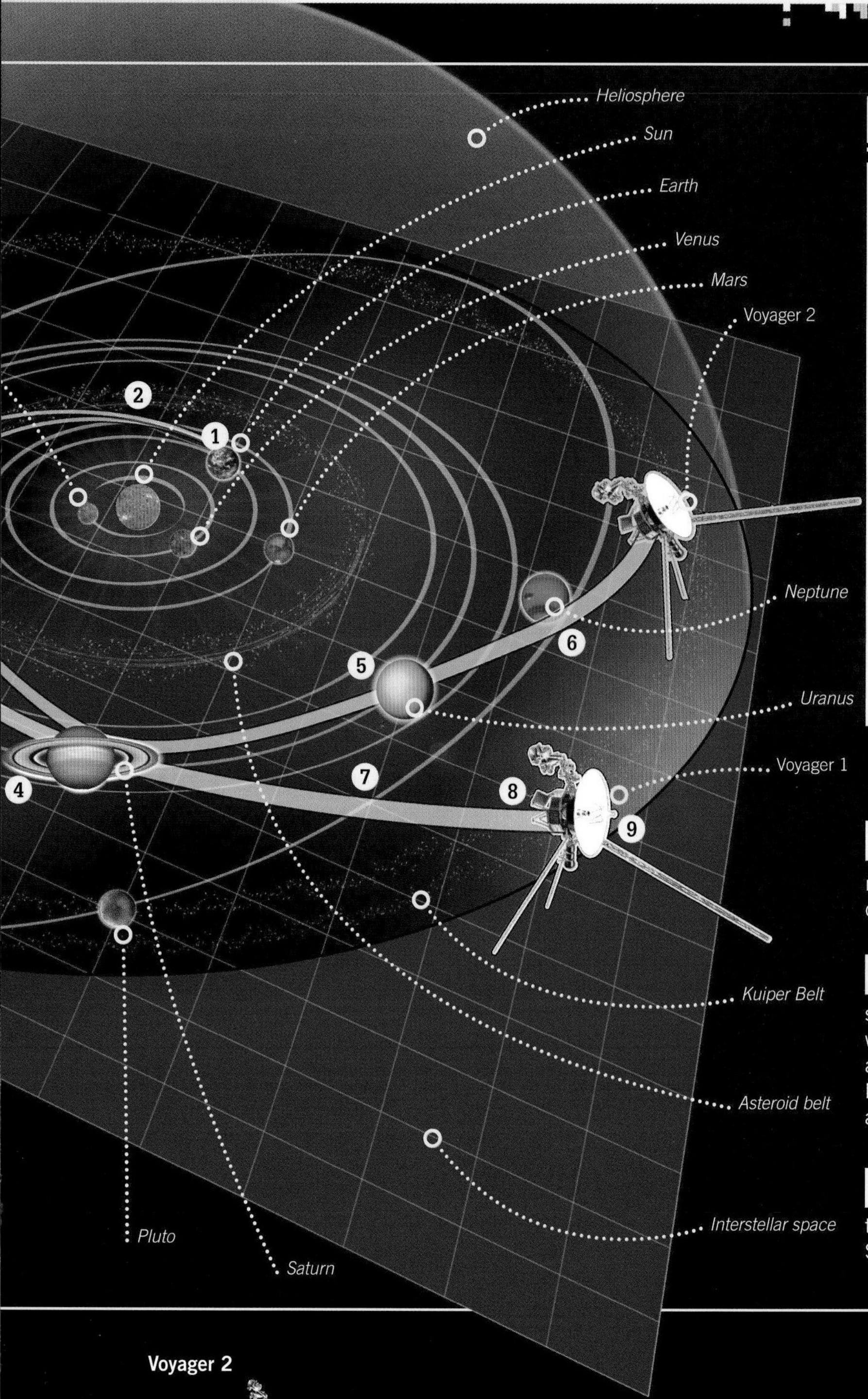

Heliosphere

Sun

Earth

Venus

Mars

Voyager 2

Neptune

Uranus

Voyager 1

Kuiper Belt

Asteroid belt

Interstellar space

Pluto

Saturn

LOOKING BACK

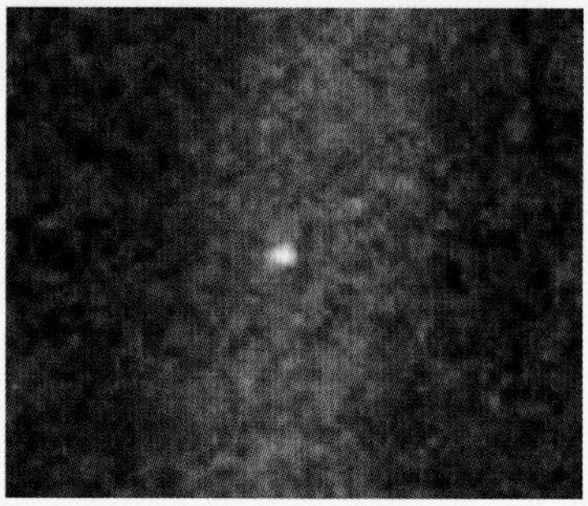

In 1990, NASA turned *Voyager 1*'s camera back towards Earth. From a distance of 6 billion km (3.7 billion miles), Earth was revealed as just a "pale blue dot" – "a mote of dust suspended on a sunbeam", as American scientist Carl Sagan described it.

7 **_Pale Blue Dot image_**
At about 6 billion km (3.7 billion miles) from Earth, *Voyager 1* snaps what is often called the "Pale Blue Dot" image.

8 **_Termination shock_**
Voyager 1 crosses the termination shock (the region of space where the solar wind drops from supersonic speeds and interacts with interstellar space) in December 2004. *Voyager 2* crosses it almost three years later in September 2007.

9 **_Bye bye_ Voyager_!_**
Voyager 1 finally passes through the heliosphere and leaves the Solar System on 12 September 2013.

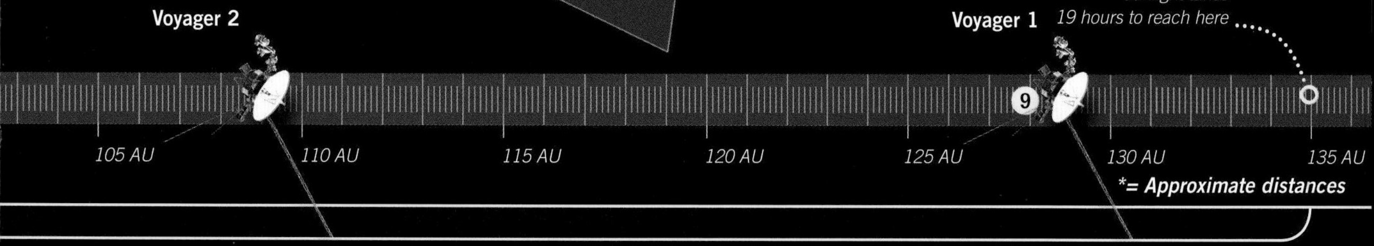

Voyager 2

Voyager 1

Sunlight takes
19 hours to reach here

105 AU   110 AU   115 AU   120 AU   125 AU   130 AU   135 AU

*= Approximate distances

IS THERE LIFE ON MARS?

AN ITALIAN ASTRONOMER turned his telescope towards Mars in 1877, and what he saw prompted speculation of **advanced Martian civilizations** that lasted almost a century. It was not until NASA's *Mariner 6* and *7* probes travelled to the Red Planet in 1969 that Mars was revealed to be the desolate "almost-Earth" we know today.

Further investigations revealed that **Mars lost hold of its atmosphere billions of years ago** – leaving only a tenuous carbon dioxide atmosphere. Then the life-hunting *Viking* landers of the 1970s probed the Martian soil and came up empty handed. It seemed that life on Mars (even on its smallest scale) would **remain the stuff of science fiction**.

Recent discoveries of **water ice around the planet's poles**, and the so-far-unexplained presence of methane in isolated regions **(alien-bacteria farts?)**, have raised the spectre of Martian life once again. In the coming years, an armada of life-hunters will be heading for the Red Planet – so here is a special look at the past, present, and future of the search for life on Mars.

1975 **1980** **1985**

1975: *Viking 1* and *Viking 2*

Launched in 1975, *Viking 1* and *Viking 2* were the first dedicated attempts to find signs of Martian life. The craft performed tests designed to detect signs of life. One test involved scooping up Martian topsoil and heating it to 500°C (932°F). *Viking* then analysed the vaporized dirt for signs of organic molecules. The test was sensitive to levels of organic compounds of only a few parts per billion. Yet no organic compounds were ever found. It seemed that the search for life had ended before it really began. In 2006, scientists replicated the *Viking* experiments with samples from Earth, and also failed to identify any organic material in any of the samples.

THE *VIKING* LANDERS WERE DESIGNED TO LAST FOR ONLY 90 DAYS ON MARS, BUT THEY BEAMED IMAGES AND DATA TO EARTH FOR MANY YEARS

Biology processor

Meteorology sensors

Radioisotope thermal generator

Sample return arm

Viking 1

VIKING

Size: 2.5 m × 2.5 m (8.2 ft × 8.2 ft)

Mass: 2,328 kg (5,132 lb); 1,445 kg (3,185 lb) of that was fuel

Arms: Up to 3 m (9.8 ft, telescopic)

Power: Two radioisotope thermal generators (RTG) containing plutonium-238

CANALS AND *CANALI*

Giovanni Schiaparelli was the Italian astronomer whose observations of Mars in 1877 led to theories about life there. Schiaparelli spotted a network of straight lines criss-crossing the Martian surface, and described them as channels, or *canali*. A mistranslation led Percival Lowell, an American astronomer, to imagine a Martian civilization whose crops were watered by a network of irrigation canals. It was not until 1969 that NASA's *Mariner* probe finally confirmed the "canals" were nothing more than optical illusions on the Martian surface.

Schiaparelli's drawing of canali *on the Martian surface*

Mars

1990 **1995**

1996: Martian fossils

On 6 August 1996, NASA announced the discovery of evidence of fossil life in a meteorite that originated from Mars and had landed on Earth 13,000 years ago. Under the scanning electron microscope, structures were revealed that seemed to be fossilized Martian bacteria. The finding remains controversial, with some scientists claiming the fossils were just the result of earthly contamination. Recent studies have found that cracks in the meteorite are filled with carbonate materials that suggest the presence of water on Mars about four billion years ago. Complex organic compounds – called polycyclic aromatic hydrocarbons – have also been identified that point to a biological origin. Interestingly these have been found deep within the rock, where contamination is unlikely.

LIFE, JIM, BUT NOT AS WE KNOW IT

Perhaps the search for life in extreme environments such as Mars should begin on Earth? All over our planet we have found life, not clinging on, but actually thriving in some of the most inhospitable environments – such as deep in Arctic ice. In 2008, the European Space Agency (ESA) sent 664 examples of these extremophiles to the International Space Station (ISS). For 18 months, two-thirds of the samples were exposed to the vacuum, massive temperature swings, and desiccating conditions of open space. The rest were exposed to a thin carbon dioxide atmosphere that simulated the Martian environment. Many of the samples survived the ordeal, with one of the stars of the show being a strange pond-dwelling creature called a tardigrade, which can survive temperature swings ranging from -272°C (-457°F) to 150°C (302°F).

Tardigrade

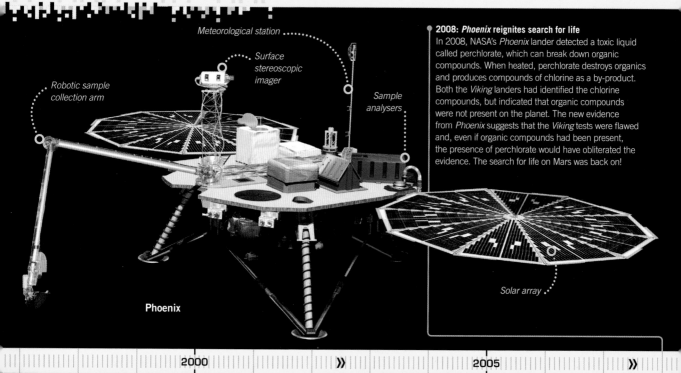

Meteorological station

Surface stereoscopic imager

Robotic sample collection arm

Sample analysers

2008: *Phoenix* reignites search for life
In 2008, NASA's *Phoenix* lander detected a toxic liquid called perchlorate, which can break down organic compounds. When heated, perchlorate destroys organics and produces compounds of chlorine as a by-product. Both the *Viking* landers had identified the chlorine compounds, but indicated that organic compounds were not present on the planet. The new evidence from *Phoenix* suggests that the *Viking* tests were flawed and, even if organic compounds had been present, the presence of perchlorate would have obliterated the evidence. The search for life on Mars was back on!

Solar array

Phoenix

2000   »   2005   »

FUTURE WEAPONS OF EXPLORATION

Ever since the first man-made craft set down on the Martian surface, Mars exploration has been dominated by static landers or lumbering rovers. In the future, however, planetary explorers might be much smaller...

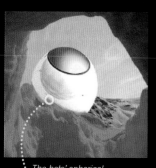

MICROBOT SWARM
These tiny spherical robots could be dropped in thousands on the Martian surface where they would operate as a swarm to locate and explore caves. Just 12 cm (5 in) wide, the individual robots use a powered leg to hop, bounce, and roll their way across difficult terrain.

The bots' spherical shape makes them suited to exploring crevices and cave systems.

Each nanobot consists of a polymer skin covering a 1 mm (0.04 in) computer chip.

The wrinkled skin allows the nanobot to be picked up by the wind.

A chip applies a small electric charge to the smooth skin, causing it to wrinkle.

NANOBOT DUST SWARM
Currently under development, these potential Martian explorers would be no larger than a grain of sand. Clouds of "smart dust" containing up to 30,000 nanobots could be dispatched into the Martian atmosphere where, taking advantage of Mars's low gravity (just 38 per cent of Earth's), they would be carried by the wind.

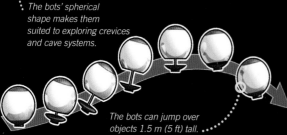

The bots can jump over objects 1.5 m (5 ft) tall.

2012: *Curiosity*

In September 2013, NASA announced that *Curiosity* had detected "abundant, easily accessible" water in the Martian soil. The robotic explorer had found that the red surface of Mars contains about two per cent water by weight – meaning that future colonists could (in theory) extract about a litre of water from every cubic foot of Martian dirt – meaning that the life on Mars could soon be us!

Curiosity

CURIOSITY

Size: 2.9 m x 2.7 m (9.5 ft x 8.9 ft)

Mass: 899 kg (1,982 lb)

Arm: 2.1 m (6.9 ft)

Power: Two radioisotope thermal generators (RTG) containing plutonium-238

2010 〉〉 2015 〉〉 2020

TETRAHEDRAL WALKERS

Developed by NASA, the Addressable Reconfigurable Technology (ALT) walker consists of intelligent nodes connected by extendable struts. Motors in the nodes are used to expand or retract the connecting struts to change the walker's shape. It would also mean that, if the walker were damaged, it would be able to fix itself by removing damaged sections and rejoining to undamaged nodes.

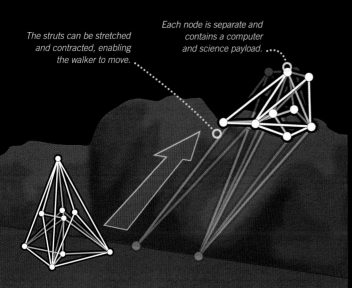

The struts can be stretched and contracted, enabling the walker to move.

Each node is separate and contains a computer and science payload.

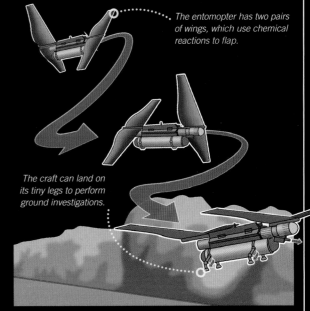

The entomopter has two pairs of wings, which use chemical reactions to flap.

The craft can land on its tiny legs to perform ground investigations.

INSECT-INSPIRED ENTOMOPTERS

Designed as part of NASA's Institute for Advanced Concepts, the entomopter is a flying robot modelled on a hawk moth. The Martian atmosphere is too thin (one per cent of Earth's) for fixed-wing aircraft, but is perfect for a lightweight flapping vehicle.

COLONIZING **MARS**

BY 1970, AMERICA had the Moon under its belt and the **human exploration of other worlds** was riding high in the imagination of the Earth-bound masses. Predictions of **lunar colonies** by the late 1970s and **Martian colonies** by the 1980s were tossed around the media as if their planning and execution were no more troublesome than building a highway.

By today, Mars was supposed to be a **"New Earth"** where humans no longer tenuously inhabited Martian outposts, but thrived in autonomous cities where generations were born, lived, and died having never known the blue skies of Mother Earth. Obviously this is not the case today, nor is it likely to be for a very long time.

However, we might soon see the dispatch of those first Martian pioneers and the settlement of those first outposts – even if they are three decades too late. In 2010, NASA announced **an initiative to move space flight and exploration to the next level.**

The plan has been dubbed the **"Hundred Year Starship"** – and that is about it (as they have not been forthcoming with many details) – and has received funding from NASA and its wacky research arm, the Defense Advanced Research Projects Agency (DARPA).

The idea is to develop a **new form of spacecraft** that would cut the journey time to Mars (currently a prohibitive six to nine months) and, arguably more importantly, cut the cost. Under discussion is a propulsion system called **"microwave thermal propulsion"**. A craft powered in such a way would have its energy "beamed" via microwaves, or laser, directly from Earth. Such beams would heat its propellant directly and push the craft forward – thus eliminating the massive amounts of fuel it would otherwise have to carry with it (which is heavy, and heavy stuff costs a lot to get off the ground).

Halving the distance that a manned craft might need to travel would also cut costs. How? Well, by making it a **one-way trip for the astronauts on board.**

The NASA proposal suggests that the best way to conquer Mars might be to land the first pioneers on the Red Planet – or initially on its moon, Phobos (see right) – and then **leave them there, forever.** That is not to say that they would be dumped and then left to fend for themselves. They would be **periodically re-supplied from Earth with basic necessities,** but otherwise, they would be encouraged to become increasingly self-sufficient. Despite the "no return" clause, NASA is not expecting to have any trouble recruiting volunteers.

PHOBOS: A PERFECT FRONTIER POST?

Measuring just 28 km (17.3 miles) wide, with just two billionths of Earth's mass, Mars's largest moon is little more than an asteroid. It has no atmosphere at all and its gravity is infinitesimally small. It is also very close to Mars – at a distance of just 9,377 km (5,826 miles) – all of which might make it a perfect Martian "jumping off" point.

WEAK GRAVITATIONAL FIELD

Using Phobos as a base camp, scientists could explore the surface of Mars with telescopes and remote-controlled rovers. And, because its gravitational field is so weak, landing is a doddle and taking off wouldn't require much energy. This would make it cheaper and easier to send spacecraft from Earth to Phobos (then ferry humans and materials down to Mars) than to send them directly to the Martian surface.

Phobos

Return vehicle

Return capsule

Flight module

Main propulsion

PHOBOS-GRUNT

In 2011, Russia launched *Phobos-Grunt* (meaning "Phobos-soil") to take samples from Phobos and return them to Earth. The mission failed, but in 2012, a repeat mission was announced – to be carried out in 2020.

COLONISTS WILL HAVE TO DEAL WITH RAZOR-SHARP DUST THAT WILL MUCK UP MACHINERY AND SPACESUITS

On Mars:
This artist's impression shows a pioneering astronaut zipping around on a scooter on the Red Planet.

WHY WE NEED TO SPEED
THE JOURNEY UP A BIT

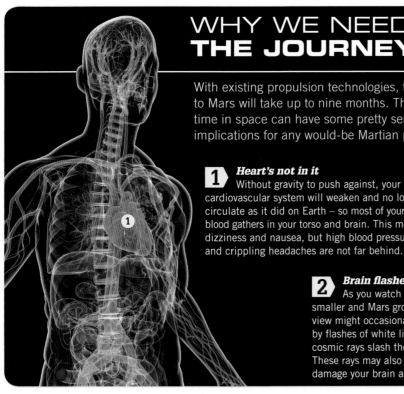

With existing propulsion technologies, the journey to Mars will take up to nine months. This amount of time in space can have some pretty serious health implications for any would-be Martian pioneer.

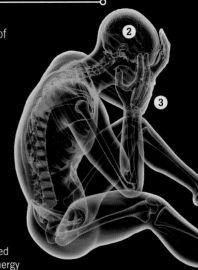

1 **Heart's not in it**
Without gravity to push against, your cardiovascular system will weaken and no longer circulate as it did on Earth – so most of your blood gathers in your torso and brain. This means dizziness and nausea, but high blood pressure and crippling headaches are not far behind.

2 **Brain flashes**
As you watch Earth become smaller and Mars grow larger, your view might occasionally be obscured by flashes of white light as high-energy cosmic rays slash though your brain. These rays may also cause cancers, and damage your brain and nervous system.

LOST IN **SPACE**

Of all the missions launched towards the Red Planet, about two-thirds have either been blighted with technical problems, or been lost completely. Here are the missions that have fallen foul of the "Curse of Mars".

Mariner 3

1960, USSR
Marsnik 1 and 2: Mars fly-by missions – both lost to launch failures.

1964, USA
Mariner 3: Mars fly-by – spacecraft housing failed to open following launch. Unable to deploy its solar panels, the craft ran out of power. It is still orbiting the Sun.

| 1960 | » | 1965 | » | 1970 | » | 1975 | » | 1980 |

Sputnik 22

1962, USSR
Sputnik 22: Mars fly-by – launch failure. Destroyed in low-Earth orbit.
Mars 1: Mars fly-by – contact lost en route to Mars.
Sputnik 24: Mars lander – destroyed in low-Earth orbit.

1969, USSR
Mars 1969A: Mars orbiter and lander – launch failure. Lost in explosion.
Mars 1969B: Mars orbiter and lander – launch failure. Lost in launchpad explosion.

1971, USA
Mariner 8: Mars orbiter – lost during launch failure.

Mariner 8

1973, USSR
Mars 4: Mars orbiter – reached Mars but failed to fire braking thrusters and the craft over-shot the planet.
Mars 6: Mars lander – reached Mars but lander was lost on descent.
Mars 7: Mars lander – reached Mars but lander separated three hours too early and over-shot the planet.

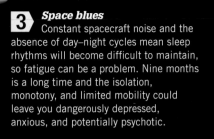

3 **Space blues**
Constant spacecraft noise and the absence of day–night cycles mean sleep rhythms will become difficult to maintain, so fatigue can be a problem. Nine months is a long time and the isolation, monotony, and limited mobility could leave you dangerously depressed, anxious, and potentially psychotic.

4 **No exercise**
With little opportunity to exercise, the muscles in your flabby, underused limbs will atrophy, making movement awkward and painful.

5 **Bad bones**
More than 200 days of near-total weightlessness causes your bones to excrete calcium and phosphorus, meaning they have lost as much density as they would during a lifetime on Earth – making your bones fragile and prone to fracture. No longer compressed by gravity, your vertebrae can separate, causing backaches.

VASIMR

Existing rocketry technologies are too slow for effective Martian colonization. Experimental technologies, such as the (awesomely named) Variable Specific Impulse Magnetoplasma Rocket (VASIMR), could reach speeds of 193,000 kph (119,925 mph) and get to Mars in 39 days. But this is still 20 to 30 years away.

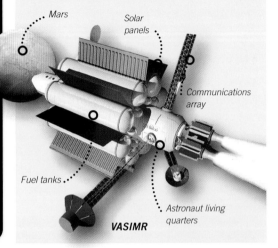

Mars

Solar panels

Communications array

Fuel tanks

Astronaut living quarters

VASIMR

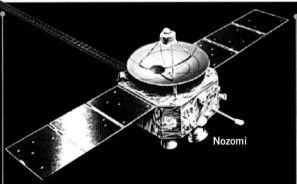

1988, USSR
Phobos 1: Mars orbiter and Phobos lander – lost en route to Mars when command failure caused steering thrusters to shut down.
Phobos 2: Mars orbiter and Phobos lander – successfully entered Mars orbit but contact was lost during attempt to deploy the landers.

Nozomi

1998, Japan
Nozomi: Mars orbiter – failed to achieve Mars orbit due to electrical failure.

1998, USA
Mars Climate Orbiter: Mars orbiter – communication problem caused craft to break up in Martian atmosphere.

1990   1995   2000   2005

1992, USA
Mars Observer: Mars orbiter – contact lost three days before reaching Mars orbit.

Phobos 2

1996, Russia
Mars 96: Mars orbiter and lander – lost during launch failure.

1999, USA
Mars Polar Lander and Deep Space 2: Mars lander and surface penetrator – lost during descent to the planet's surface.

2003, Britain
Beagle 2: Mars lander – lost during descent to the planet's surface.

MAPPING THE
MILKY WAY

IN 1676, THE ENGLISH ASTRONOMER ROYAL, John Flamsteed, sat down to compile the first catalogue of star positions to be recorded with the aid of a telescope. He spent **an incredible 43 years dedicated to the task** and, by the time the final catalogue was published in 1725 (six years after his death), he had **recorded the positions of nearly 3,000 stars with unrivalled precision.**

Nearly 300 years later, a mission was launched with the aim of mapping the positions of **one billion stars** with a level of accuracy that would have made Flamsteed's head explode – and it plans to do so **in just five years.**

Launched on 20 December 2013, the European Space Agency's (ESA's) Gaia spacecraft was one of the **most ambitious space-charting missions ever conceived.** From its position 1.5 million km (932,000 miles) from Earth, Gaia will map the **precise location, composition, brightness, and age** of a billion stars to create the ultimate three-dimensional map of our corner of the Milky Way.

Mapping space:
Gaia spins slowly, sweeping its two telescopes across the entire sky. The light from millions of stars is focused simultaneously on to a camera sensor that is the largest ever flown into space.

GAIA PINPOINTS THE POSITION OF STARS WITH AN ACCURACY A HUNDRED TIMES GREATER THAN ANY EXISTING STAR CATALOGUE

For a lucky 150 million of those stars, Gaia aims to chart how they are moving through space. Their exact speed through the galactic medium will be measured as well as their motion relative to Earth – building a three-dimensional map that will allow astronomers to **trace the origins and evolution of the Milky Way and even provide clues about its ultimate fate.**

As if this was not ambitious enough, Gaia's remarkable near-billion-pixel camera will simultaneously **map the locations of thousands of asteroids, comets, planetary systems, supernovae, and even distant galaxies.**

Gaia is armed with two telescopes that will sweep the sky to a depth of 20,000 parsecs, or 65,200 light-years – generating so much data that it will take the number-crunching power of a supercomputer to process it.

The level of precision needed to make these measurements requires absolute stability, so the craft has no moving parts and has a skeleton made of silicon carbide, which does not expand or contract when the temperature fluctuates.

SEEING STARS FROM
DIFFERENT ANGLES

There are no road signs or handy cosmic-scale tape measures in space, so astronomers have had to develop clever techniques to measure distance...

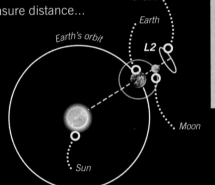

Position of Gaia

Earth

Earth's orbit

L2

Moon

Sun

STAYING STILL
Gaia sits in an area of space called a Lagrange point. The spacecraft occupies Lagrange 2 (L2), which is located beyond the Moon's orbit away from the Sun. Here the Sun's gravity and Earth's cancel each other out – allowing Gaia to remain relatively stationary.

GAIA MEASURES THE POSITIONS AND MOVEMENT OF UP TO 8,000 STARS EVERY SECOND, TO AN ACCURACY EQUIVALENT TO A COIN SITTING ON THE MOON'S SURFACE

THE PARALLAX EFFECT
Gaia will take advantage of the "parallax effect" to measure the distance to stars. Close one eye and hold your finger in front of your face and note where it appears relative to the background. If you swap eyes, you will see the finger jump to the side, even though it hasn't moved. This is the parallax effect, and it happens because your eyes see things from slightly different angles. Gaia takes readings at different positions, and combines them to find the correct distance.

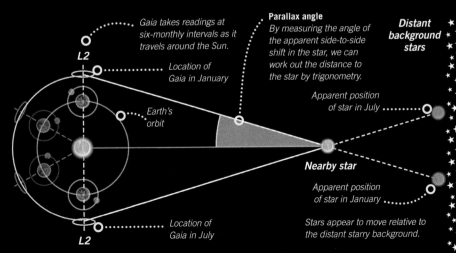

Gaia takes readings at six-monthly intervals as it travels around the Sun.

L2

Location of Gaia in January

Earth's orbit

L2

Location of Gaia in July

Parallax angle
By measuring the angle of the apparent side-to-side shift in the star, we can work out the distance to the star by trigonometry.

Distant background stars

Apparent position of star in July

Nearby star

Apparent position of star in January

Stars appear to move relative to the distant starry background.

DETECTING KILLER ASTEROIDS

GENERALS HAVE KNOWN for millennia that to prevail in battle, you must study your enemy – as the great Chinese military tactician Sun Tzu wrote in the 6th century BCE: **"know thy enemy and know thy self, you can win a hundred battles".** It is a lesson that has been put to use on battlefields all over the world, but now it is time to move the lesson to a larger field of battle: space itself.

As the meteor that exploded over the Russian region of Chelyabinsk Oblast in 2013 reminded us, the pale blue dot we call Earth is really rather small and vulnerable. That meteor **injured some 1,500 people and damaged more than 4,300 buildings across six cities.** The damage was impressive, but more impressive was the size of the offending space rock – it measured a mere 30 m (98 ft) across – **a speck of dust compared to some of the asteroids hoofing about in the space above our heads.**

So what can we do to prevent this happening again? Well, in the case of the Russian meteor, probably not much – **it was just too small to spot before it entered the atmosphere** – but we might be able to do something about the larger rocks that we can detect.

Studies suggest that there are some 4,700 near-Earth asteroids

EARTH IS STRUCK BY AN ASTEROID THE SIZE OF A FOOTBALL FIELD APPROXIMATELY EVERY 2,000 YEARS

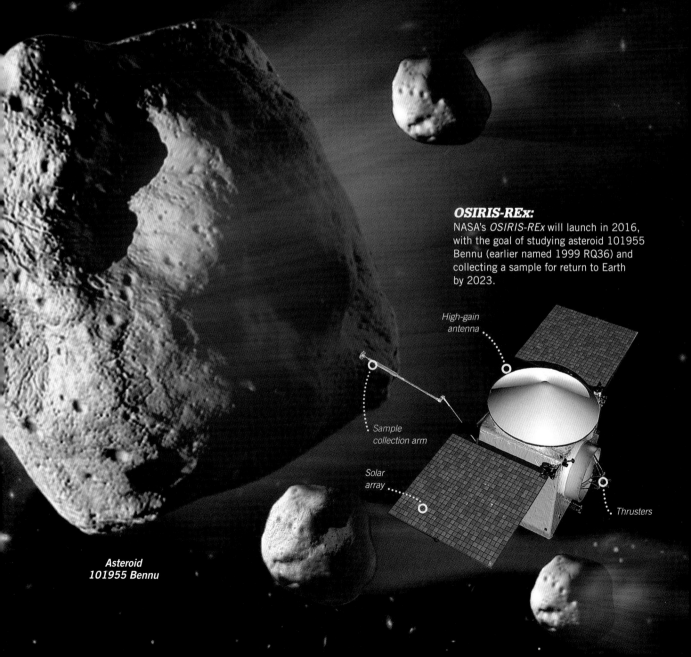

measuring in at more than 100 m (320 ft) and, although none are expected to hit Earth in the next 100 years, it would be folly not to prepare for the worst. Detection is the first line of defence because, once we know where they are, we can predict their orbits decades in advance – giving us lots of time to rally our forces.

However, to defend effectively against a potential killer asteroid, **we must first know what they are made of and how they work.**

Two new "know thy enemy" missions have been announced by space agencies on both sides of the pond. On the American side, NASA will have the grandly named *OSIRIS-REx* – which will return an asteroid sample to Earth and, on the European side, ESA will have the slightly geriatrically named AIDA, which will study the effects of crashing a spacecraft into an asteroid.

Once we figure out our foe, **how do we go about protecting ourselves from a supersonic lump of rock the size of a mountain?** Well, there are a few ideas...

OSIRIS-REx:
NASA's *OSIRIS-REx* will launch in 2016, with the goal of studying asteroid 101955 Bennu (earlier named 1999 RQ36) and collecting a sample for return to Earth by 2023.

High-gain antenna

Sample collection arm

Solar array

Thrusters

Asteroid 101955 Bennu

HOW TO DEFLECT
A KILLER ASTEROID...

In disaster movies, fictional scientists have come up with all sorts of imaginative ways to prevent asteroid-induced Armageddon. It turns out that real life solutions are even more bizarre...

A nearby nuclear explosion would heat one side of the rock, causing material to vaporize.

NUKE IT
Using nuclear weapons to save Earth from asteroid Armageddon is almost a cliché, but a direct hit would probably only serve to break the asteroid into many deadly chunks. A better option might be to detonate the warhead near the asteroid. Of course, this would require a little forward planning.

SPLAT!

GET PUSHY
If you have ever watched a tiny tugboat manoeuvre a huge ship into harbour, you might see the merit of the next idea: use a spacecraft to push the asteroid away. It really is as simple as it sounds (well, sort of). All you need to do is fly a spacecraft equipped with ion thrusters to the asteroid and push the space rock into a nice, safe trajectory. In 15–20 years, you'll be totally safe!

PEPPER IT WITH PAINTBALLS
An alternative to the solar sail idea is to use a spacecraft to blast the offending asteroid with five tonnes of paintballs. This would coat the rock in a layer of paint – providing a reflective surface that light would bounce off, changing the asteroid's trajectory.

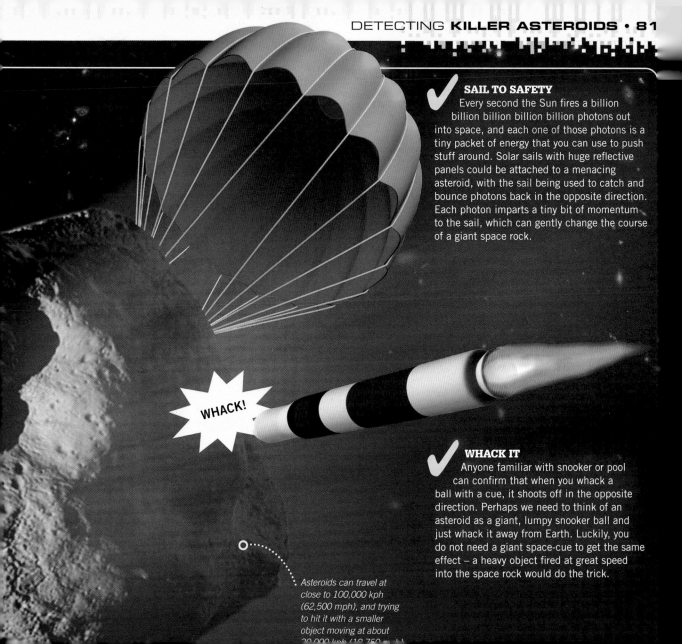

SAIL TO SAFETY

Every second the Sun fires a billion billion billion billion billion photons out into space, and each one of those photons is a tiny packet of energy that you can use to push stuff around. Solar sails with huge reflective panels could be attached to a menacing asteroid, with the sail being used to catch and bounce photons back in the opposite direction. Each photon imparts a tiny bit of momentum to the sail, which can gently change the course of a giant space rock.

WHACK!

WHACK IT

Anyone familiar with snooker or pool can confirm that when you whack a ball with a cue, it shoots off in the opposite direction. Perhaps we need to think of an asteroid as a giant, lumpy snooker ball and just whack it away from Earth. Luckily, you do not need a giant space-cue to get the same effect – a heavy object fired at great speed into the space rock would do the trick.

Asteroids can travel at close to 100,000 kph (62,500 mph), and trying to hit it with a smaller object moving at about 20,000 kph (12,750 mph)

LOOKING BEYOND MARS FOR LIFE

MARS DOMINATES the search for life beyond Earth, but a growing number of scientists believe that our efforts should be directed towards a world that seems **a most unlikely candidate for extraterrestrial life** – Enceladus, the sixth-largest moon of Saturn.

For life (as we know it) to evolve and survive, it requires **three essential ingredients – water, energy, and organic chemicals.** But how can a tiny frozen moon so far from the Sun possibly possess any of these ingredients?

Ingredient one: Water

Enceladus's northern hemisphere is heavily cratered and looks like any other moon, but its southern hemisphere is a little bit special. It is almost completely bereft of craters, which means that **the surface must be undergoing constant change.** Its cracked and scarred surface is riven by colossal canyons. Directly over the south pole are Enceladus's famous **"tiger stripes"** – four massive tears in the icy surface more than 140 km (85 miles) long and hundreds of metres deep that resemble tectonic fault lines on Earth.

In 2005, scientists working on NASA's *Cassini* mission discovered **vast plumes of water being vented from the tiger stripes** – like giant frozen volcanoes spewing ice instead of molten rock. The ice geysers of Enceladus (there

Titan, Saturn's largest moon

Saturn's rings

are more than 100 of them) burst through the moon's frozen surface at 1,300 kph (800 mph) and blast 200 kg (440 lb) of water vapour thousands of kilometres into space every second. As it

encounters the frigid vacuum of space, the liquid water instantly freezes into tiny ice crystals. Much of this falls back to Enceladus as snow, which accumulates over millions of years to form

snow drifts up to 100 m (328 ft) deep and gives the moon a white surface that reflects almost all of the Sun's feeble rays back into space – making Enceladus the most reflective object in the Solar System. But not all of it falls as snow; **some of the ice spreads out into space and wraps around Saturn, forming the planet's great E Ring.** More relevant for our recipe for life is the fact that the geysers would not be possible unless there were **liquid water beneath Enceladus's icy surface.**

Cratered northern hemisphere

ICY PLUMES

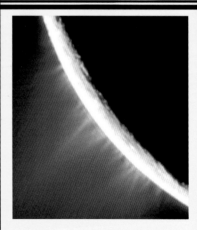

So far, astronomers have found 101 geysers on Enceladus venting water vapour and ice from near the moon's south pole. *Cassini* took this spectacular shot of the ice plumes, backlit by the Sun.

Enceladus

Tiger stripe

Ingredient two: Energy

Being too small to generate its own internal heat and so far from the warmth of the Sun, **Enceladus should be frozen solid.** What was causing that ice to melt? When *Cassini* investigated Enceladus's "tiger stripes" with its thermal imaging cameras, it discovered that the stripes sat over **"hot spots" that are much warmer than the rest of the moon.** Heat is relative, of course. These "hot" zones are about -95°C (-139°F), which on Earth is pretty cold, but 1.4 billion km (869 billion miles) from the Sun and on a moon with an average temperature of -200°C (-328°F), it is almost balmy.

But where is the energy coming from? At such a great distance from the Sun, you can be sure it is not coming from there. The answer lies in that most enigmatic of forces: gravity. Any object that orbits another exerts a gravitational influence on that object called the tidal force. Earth pulls on the Moon and the Moon pulls on

Earth. Although much smaller than Earth, **the Moon's gravitational pull is strong enough to distort Earth's solid rock crust** – lifting it by 20 cm (7.8 in) with every pass. If a relatively small object like the Moon can distort Earth, just imagine what Saturn does to tiny Enceladus. Saturn is about 95 times more massive than Earth and Enceladus is only one-sixth the size of the Moon.

Enceladus travels around Saturn in an elliptical orbit, which means that, as the moon moves closer to and further away from Saturn, **tidal forces are continually stretching and squashing Enceladus like a ball of putty.** This constant internal movement creates friction that, as anyone who has ever enjoyed a carpet burn can attest, generates energy in the form of heat. **This heat is enough to melt Enceladus's icy interior** and create a small underground ocean of liquid water.

Ingredient three: Organic compounds

Some of *Cassini's* visits to Enceladus pass as close as 21 km (13 miles) to the moon's surface. This has enabled *Cassini* to **fly through the heart of the plumes and analyse the gas and ice.** In the process it detected a cocktail of organic compounds – ammonia, methane, carbon dioxide, acetylene, and other hydrocarbons. All these ingredients are thought to have made up the prebiotic soup from which life eventually emerged

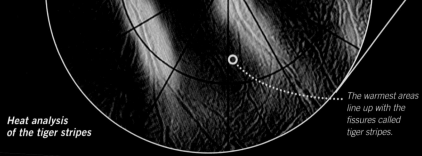

Heat analysis of the tiger stripes

The warmest areas line up with the fissures called tiger stripes.

on Earth. **But could life have formed in a frigid subterranean ocean that is so distant from our life-nurturing Sun?** It depends on your expectations of what alien life will be. If you are hoping for extraterrestrial dinosaurs or cuddly, exponentially replicating fur balls then you will be sorely disappointed. But **if you lower your ET aspirations to include microbial life, then you are in with a shot.** There are several microbial ecosystems discovered on Earth that could provide a blueprint for possible Enceladian life. One group of singled-celled life, called archaea, are able to thrive in the most extreme of environments. Known as methanogens (because they give off methane as a by-product of their metabolism), they have been **found living locked away from oxygen and sunlight under kilometres of ice in Greenland.** Such microbes have been discovered surviving on the energy from the chemical interaction between different kinds of minerals and even living off the energy produced by radioactive decay in rocks.

THE TIGER STRIPES ARE EMITTING AN ENORMOUS AMOUNT OF ENERGY – ABOUT 16 GIGAWATTS, WHICH EQUATES TO ABOUT 20 COAL-FIRED POWER STATIONS

So Enceladus has the three basic ingredients in the recipe for life – water, energy (warmth), and organic chemicals – and we know that microbial life can survive just about anywhere, but does life exist below the moon's cue-ball surface? Only a dedicated sampling mission can hope to answer this question, but it is an intriguing thought, is it not?

HOW TIDAL FORCES
COULD CREATE AN OCEAN ON ENCELADUS

The interior of Enceladus should be frozen solid, so what is melting the ice? The answer lies with its giant companion, Saturn, and the moon's elliptical orbit.

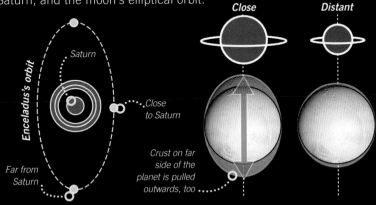

1 Elliptical orbit
Enceladus moves in an elliptical orbit around Saturn.

2 Distance from Saturn
The closer Enceladus's orbit carries it to Saturn, the more the moon is distorted by the gas giant's gravity.

3 Wobble wobble
It is thought Enceladus also wobbles slightly as it orbits (called libration) – meaning the moon is stretched and pulled in different directions.

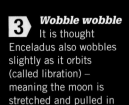

4 Stretching out
All this stretching creates friction within Enceladus – generating heat, which melts the ice and creates a subsurface ocean. The water from this ocean travels up through the ice and collect in caverns beneath the tiger stripes before venting out into space.

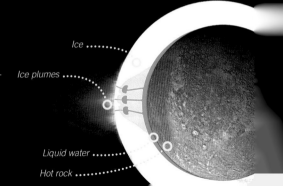

A WEBB TO CATCH
THE OLDEST STARS

IN THE KINGDOM OF TELESCOPES, there is no denying that the Hubble Space Telescope is king. From the moment of its launch (well, from the moment its faulty optics were fixed with the addition of a set of space spectacles), it has beamed back images that have **revolutionized our understanding of the cosmos**, tapped into humanity's collective imagination, populated the coffee tables of the world with countless pictorial tomes, and titillated the planet's computer-users with an endless stream of astro-screensavers. But even legends must one day step aside and cede their title to the next generation. **Hubble's heir-apparent is the James Webb Space Telescope** (JWST).

But transitions of power rarely run smoothly, and **Webb's ascension has certainly not gone to plan**. Since it was first conceived in 1996, it has been delayed (from an initial launch date of 2013 to 2018), run over budget by billions of pounds – from £2.2 billion to £5.6 billion ($3.6 billion to $9.1 billion) – and **even been cancelled**, but it seems, at last, that Webb is finally on course to assume its heavenly throne.

In 2013, the European Space Agency (ESA) announced it had completed the second of the two instruments it is contributing to NASA's mighty orbiting observatory. Called the Near-InfraRed Spectrograph, or NIRSpec, it is an infrared camera that will be **sensitive enough to detect light that has been travelling across space for 13.6 billion years** – revealing the very first stars and galaxies to flare into life just 400 million years after the Big Bang. It follows hot on the heels of Europe's other contribution – the British-designed and built Mid-Infrared Instrument (MIRI).

Studying the Universe in infrared will also allow astronomers to pierce the opaque gloom of cosmic dust that obscures so much of the cosmos, **revealing objects that Hubble is simply blind to**. NIRSpec's spectrometer will give Webb the power to study the atmospheres of distant worlds and discern their chemical composition – **a first step towards finding alien life**.

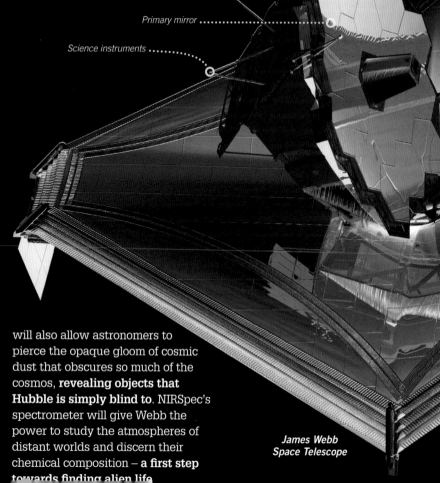

Primary mirror

Science instruments

James Webb Space Telescope

JWST VS HUBBLE

Like Hubble, Webb will see visible light, but its real talent will be capturing infrared light using its enormous 6.5 metre (21.3 ft) primary mirror.

THE ORBIT

Webb will not orbit close to Earth, like Hubble – instead, it will inhabit an area in space called a Lagrange point. Webb will occupy Lagrange 2, which is a region about 1.5 million km (930,000 miles) from Earth, where the Sun's gravity and Earth's gravity cancel each other out – allowing the craft to remain relatively stationary.

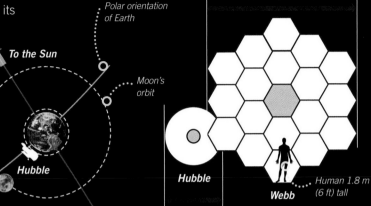

Polar orientation of Earth

To the Sun

Moon's orbit

Hubble

Webb will rotate around the L2 point in a "halo" orbit.

L2   **JWST**

Secondary mirror

6.5 m (21.3 ft)

Hubble

2.4 m (7.8 ft)

Webb

Human 1.8 m (6 ft) tall

THE MIRROR

Webb's patchwork of hexagonal mirrors has about seven times the light-collecting area of Hubble's, and has a field of view more than 15 times larger.

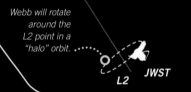

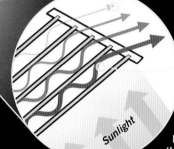

Very little heat makes it through to the telescope.

Each layer blocks and deflects some heat. The deflected heat vents away from the telescope.

Sunlight

Working of the sunshield

Webb is a very different beast from Hubble. More of an **interstellar sailing ship** than a telescope, Webb's colossal light-collecting mirrors sit atop a sunshield the size of a tennis court and, once unfolded to their full spread, the **array of hexagonal mirrors will dwarf Hubble's single mirror.**

Nor will Webb have access to the sort of home comforts that Hubble has enjoyed in its Earth-hugging orbit. Webb will be well beyond the reach of the servicing missions that have repaired and upgraded Hubble, and, should anything go wrong, **it will be well beyond any sort of help at all**. They do say it is lonely at the top.

JWST'S GOALS

• Search for light from the first stars and galaxies to form after the Big Bang.
• Study galaxy formation and evolution.
• Study planetary systems and the origins of life.

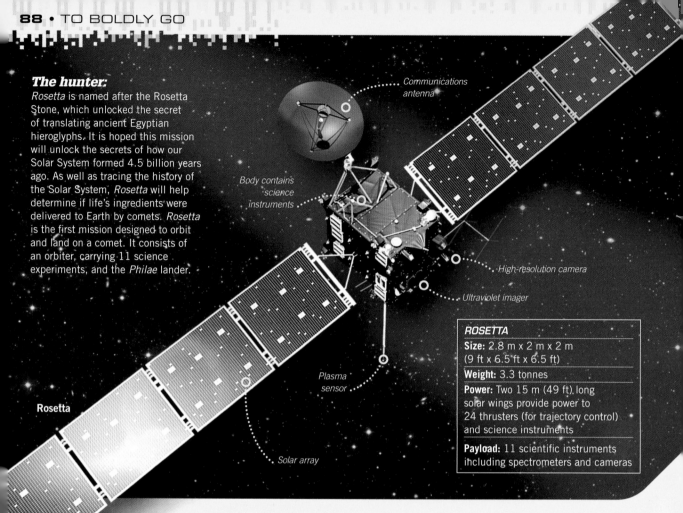

The hunter:
Rosetta is named after the Rosetta Stone, which unlocked the secret of translating ancient Egyptian hieroglyphs. It is hoped this mission will unlock the secrets of how our Solar System formed 4.5 billion years ago. As well as tracing the history of the Solar System, *Rosetta* will help determine if life's ingredients were delivered to Earth by comets. *Rosetta* is the first mission designed to orbit and land on a comet. It consists of an orbiter, carrying 11 science experiments, and the *Philae* lander.

Communications antenna

Body contains science instruments

High-resolution camera

Ultraviolet imager

Rosetta

Plasma sensor

Solar array

ROSETTA

Size: 2.8 m x 2 m x 2 m (9 ft x 6.5 ft x 6.5 ft)

Weight: 3.3 tonnes

Power: Two 15 m (49 ft) long solar wings provide power to 24 thrusters (for trajectory control) and science instruments

Payload: 11 scientific instruments including spectrometers and cameras

ESA'S *ROSETTA* COMET CHASER

SOMEWHERE IN THE FRIGID BLACKNESS of deep space, a hunter is preparing to be stirred from her slumber. She has spent **ten long years chasing down her prey** and now, after a journey of almost 7 billion km (4.35 billion miles), she is on the brink of ensnaring her quarry. When *Rosetta* set out, she was **hopelessly out-paced by her prey**, but, after four tours of the inner Solar System (stealing gravitational energy from the planets she encountered along the way) her speed of more than 135,000 kph (83,885 mph) is more than a match for the object in her sights.

But the hunt had been exhausting and, millions of kilometres from home, the Sun had been too weak to sustain her, so, **for about two-and-a-half years she was hibernating** – rationing her reserves for the final pursuit.

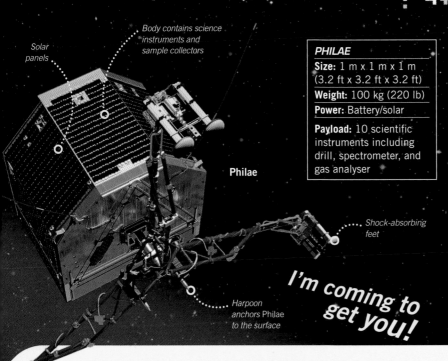

Solar panels

Body contains science instruments and sample collectors

Philae

Shock-absorbing feet

Harpoon anchors Philae to the surface

I'm coming to get you!

PHILAE	
Size: 1 m x 1 m x 1 m (3.2 ft x 3.2 ft x 3.2 ft)	
Weight: 100 kg (220 lb)	
Power: Battery/solar	
Payload: 10 scientific instruments including drill, spectrometer, and gas analyser	

The hound:

The lander is named after the island Philae in the river Nile, where an obelisk was found that helped decipher the Rosetta Stone. The *Philae* lander is the first spacecraft to make a soft landing on the surface of a comet. It piggybacks along with *Rosetta* until it arrives at the comet – where it ejects, unfolds its legs, and descends. On landing, *Philae* anchors itself to the surface by firing a harpoon into the surface. A small drill allows *Philae* to take samples, which are analysed to determine the comet's composition.

On 20 January 2014, *Rosetta*'s handlers at the European Space Agency (ESA) sent a signal to *Rosetta* that sparked up circuits, turned on heaters, and triggered instruments – **waking the hunter at precisely 10 am.**

After two-and-a-half years sleeping in the freezer, it took some time for *Rosetta*'s instruments to wake up fully and send a message to Earth. There were several tense hours before her operators knew that the huntress had survived her hibernation. *Rosetta* then began a series of manoeuvres that, over time, **saw her fall into line behind the comet, and eventually catch up in August 2014.** She gradually entered into an orbit around the 5 km (3 mile) wide lump of rock and ice. Once there, she mapped the surface of Comet 67P/Churyumov-Gerasimenko and unleashed her "hound", *Philae*. *Philae* was armed with a harpoon that it used **to spear the comet and secure itself to the surface.** The probe then deployed a drill to extract samples of the comet to be studied by its panoply of scientific instruments.

Sunlit side

The quarry:

Rosetta's target is Comet Churyumov-Gerasimenko. A remnant of the formation of the Solar System, this 4 km (2.5 mile) wide lump of rock and ice was discovered in 1969, but has been knocking around for 4.5 billion years. By the time *Rosetta* caught up with Churyumov-Gerasimenko, the comet was some 600 million km (372 million miles) from the Sun and its nucleus was quite dormant, but as it approaches the Sun, the comet will warm up and its ices will "boil" off (sublimate) – forming the comet's trademark tails.

Comet Churyumov-Gerasimenko

THE **CHASE...**

By the time Rosetta caught up with comet Churyumov-Gerasimenko, the craft had completed almost five circuits of the inner Solar System and covered a distance of more than 6,500 million km (4,038 million miles). Along the way, the craft used the gravitational pull of Earth and Mars to accelerate from its launch speed of 26,000 kph (16,155 mph) to the 135,000 kph (83,885 mph) it needed to chase down the comet.

14 *Towards the Sun*
In December 2014, *Rosetta* accompanies the comet as it travels towards the Sun. As the comet warms, its ices "sublimate" (pass straight from solid to gas) and are ejected at supersonic speeds. *Rosetta* records and studies these changes.

1 *Blast off!*
Rosetta launches from Kourou, French Guiana, onboard an Ariane 5 rocket in March 2004.

2 *Earth slingshot 1*
A year after launch, the craft uses Earth's gravity to accelerate.

3 *Mars slingshot 1*
February 2007.

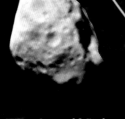

4 *Earth slingshot 2*
November 2007.

5 *Asteroid Steins*
The craft passes within 800 km (497 miles) of the 5 km (3 mile) wide asteroid, and collects information and images in September 2008.

6 *Earth slingshot 3*
The final Earth slingshot happens in November 2009.

7 *Asteroid Lutetia*
In July 2010, *Rosetta* passes the 100 km (62 mile) wide asteroid Lutetia at a distance of about 5,000 km (3,106 miles).

Sun

Earth

1st orbit

4th orbit

5th orbit

Asteroid Lutetia

Comet's orbit

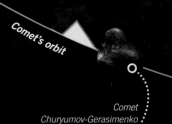

Comet Churyumov-Gerasimenko

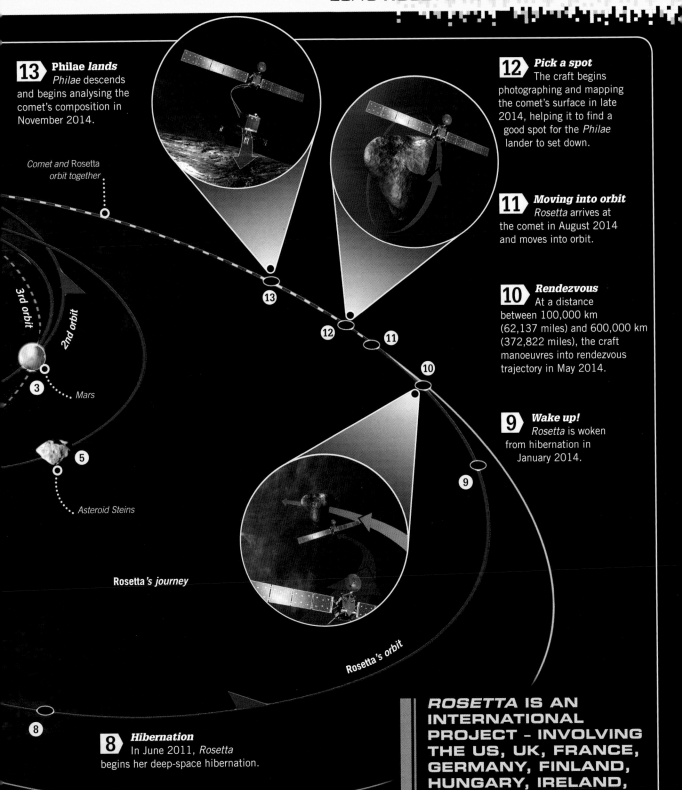

13 **Philae** *lands*
Philae descends and begins analysing the comet's composition in November 2014.

Comet and Rosetta *orbit together*

12 *Pick a spot*
The craft begins photographing and mapping the comet's surface in late 2014, helping it to find a good spot for the *Philae* lander to set down.

11 *Moving into orbit*
Rosetta arrives at the comet in August 2014 and moves into orbit.

10 *Rendezvous*
At a distance between 100,000 km (62,137 miles) and 600,000 km (372,822 miles), the craft manoeuvres into rendezvous trajectory in May 2014.

3rd orbit

2nd orbit

3

: *Mars*

5

: *Asteroid Steins*

9 *Wake up!*
Rosetta is woken from hibernation in January 2014.

Rosetta *'s journey*

Rosetta's orbit

8

8 *Hibernation*
In June 2011, *Rosetta* begins her deep-space hibernation.

ROSETTA IS AN INTERNATIONAL PROJECT – INVOLVING THE US, UK, FRANCE, GERMANY, FINLAND, HUNGARY, IRELAND, ITALY, AND AUSTRIA

GRAVITY LENSING TO SEE THE COSMOS

IT MIGHT SEEM COUNTERINTUITIVE,

but the best way to see an object in the most distant recesses of the cosmos is to make sure that you have a nice big galaxy nearby that **completely blocks your view**. Confused? Well, in the weird world of astronomy, not only can you see a distant object parked squarely behind a massive galaxy, but you can see it **bigger and brighter** than you could by using even the very best of telescopes. The phenomenon, known as gravitational lensing, takes advantage of the fact that massive objects warp the fabric of the Universe in such a way that light from a distant object is actually bent around the obscuring object, and focused on the other side – much **like light passing through a lens**.

Unlike a conventional lens, a gravitational lens **creates multiple images of the same object from different angles**. This is because the galaxy's gravity pulls in light from angles that would have seen it travel elsewhere in space. These multiple clues provide far more information than a single observation could. By understanding how long it took the light that makes up each image to travel along each path and its speed, researchers could calculate how far away the galaxy is, the overall scale of the Universe, and how it is expanding.

Gravitational lensing also works at smaller scales and can be used for planet-hunting. A nearby star, for example, can be used to infer the presence of planets orbiting stars too distant to capture directly and even allows astronomers to figure out the planet's mass. **The results seem to confirm that dark energy (much hated by many astronomers) exists, and that it is accelerating the Universe's expansion.**

WHAT IS **GRAVITY?**

Imagine the Universe to be like a bed sheet. If you place a bowling ball on the sheet, it will make a depression in the fabric. If you roll marbles along the sheet, they will roll into the depression made by the bowling ball. The exact same thing happens in the Universe. A heavy object such as a star, or galaxy, bends the fabric of the Universe (known as spacetime). All lighter objects travelling through spacetime will be drawn into (or towards) the depression caused by the mass of that object. This is gravity.

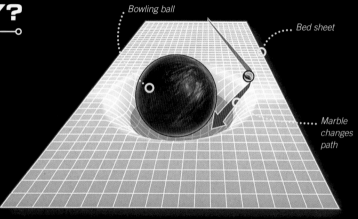

Bowling ball

Bed sheet

Marble changes path

BED SHEET UNIVERSE
The gravitational pull of stars and galaxies distorts the Universe, just as this bowling ball distorts this sheet.

USING A **GALAXY AS A SPY GLASS**...

It is not just stuff like planets (or marbles) that feel the effects of gravity – even light finds itself drawn towards massive objects. If the object is massive enough, it can act like a lens – light rays are bent around it, gathered, and focused – a phenomenon known as gravitational lensing.

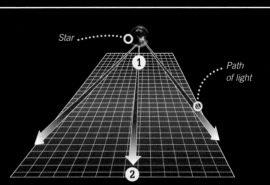

Star

Path of light

1 Even spread
Light leaves a distant astronomical object, such as a star or quasar or galaxy. In a clear and uncluttered universe, this light will spread out evenly.

2 Getting dimmer
By the time it reaches an observer on the far side of this universe, the light is so diminished that we see only a very dim image of the original object.

3 Gravity lens
When a large object nearby, such as another galaxy, blocks a distant object, you can use the galaxy's gravitational pull as a lens.

4 Warped space
The mass of the nearby galaxy warps spacetime in the galaxy's vicinity.

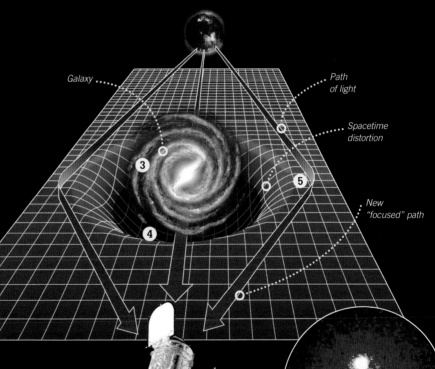

Galaxy

Path of light

Spacetime distortion

New "focused" path

5 Focusing light
Light travelling from the distant star "falls" towards the distortion created by the nearby galaxy. This pulls in light that would otherwise have spread out, bending it around the galaxy – in effect "focusing" the light on the other side.

6 Final image
This means the observer (in this case, the Hubble Space Telescope) sees a much brighter image of the distant star as more light reaches it. This image is distorted, however, as the rays show the image from different angles.

Hubble Space Telescope

Einstein cross

7 Einstein cross
A spectacular example of one sort of gravitational lensing resulted in an "Einstein cross" – where a quasar (centre) and four images of it can be seen.

ENGAGE
WARP DRIVE!

IN THE 20TH AND 21ST CENTURIES, when a traveller wanted to traverse the country in comfort and style, they took their trusty camper van. Powered by a 1.6-litre air-cooled engine, it transported its occupants on hundreds of road trips. But **by the 23rd century,** travellers were no longer content with highways and boring food and were aiming for the stars.

They soon realized that the van's trusty petrol engine was not up to the job **(it would take millions of years just to reach the nearest star),** so someone invented the **"warp drive",** fitted it to their ride, and the "interstellar" camper van was born. We are all familiar with *Star Trek*'s interstellar ship, the USS *Enterprise*, which allows Kirk and Spock to zip between stars, but surely warp-driven campers are **also the stuff of science fiction?**

If you want to travel between different star systems (without dying of old age along the way) **you need to move faster than the speed of light.** Unfortunately, Einstein proved that you cannot do this because the energy required eventually becomes infinite.

In 1994, a Mexican theoretical physicist called Miguel Alcubierre came up with **a theoretical faster-than-light propulsion method called the Alcubierre drive.** Using the Alcubierre drive, the fabric of space is manipulated to expand behind the camper and contract in front of it. Safe inside a bubble of stationary space, the camper van would be able to traverse the Universe at faster-than-light speeds **without ever physically moving.** Everything was looking good for the future of the interstellar camper van (leaving aside the practicalities of "warping" space and time).

Then scientists used supercomputers to simulate a faster-than-light journey made using an Alcubierre drive and came to a disturbing conclusion: when the camper van stops, **it will destroy everything at its destination.**

Of course, the fact that a warp drive could turn out to be the ultimate doomsday device is the least of its problems. Alcubierre's original design called for the creation of a "negative energy" bubble that **would distort the fabric of spacetime around it** – just as a massive object like Earth does, but in a much more extreme fashion. Unfortunately, an impossible amount of energy would be required to make such a bubble. It would also require some sort of "exotic matter" (which exists only in theory and, by definition, violates the laws of physics) in an amount equal to **ten billion times the total mass of the observable Universe** – that is a lot of matter, by the way. In theory, you can get round this by taking advantage of an area of physics called string theory, but it will probably be some time until you can swap your petrol engine for a warp drive.

WARPER VAN

HOW A WARP DRIVE MIGHT WORK...

String theory predicts that there are many more dimensions – perhaps as many as 26 – than the four we are familiar with. If true, it might be possible to manipulate these extra dimensions to bend space and time at will.

1 *Bubble bobble*
Around our interstellar camper van, we create a warp bubble of stationary space.

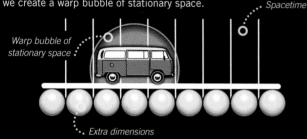

Spacetime

Warp bubble of stationary space

Extra dimensions

2 *Squashing space*
By squashing the dimensions in front of the bubble, and expanding them behind it, the camper van will be carried though space, as if on a wave.

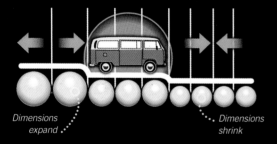

Dimensions expand

Dimensions shrink

3 *Comfortable ride*
Inside the van, the passengers are not subjected to massive acceleration forces and the van does not violate any fundamental laws of physics.

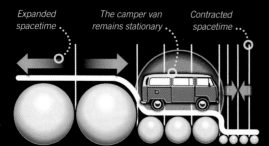

Expanded spacetime

The camper van remains stationary

Contracted spacetime

... OR PERHAPS DESTROY YOU

Warp travel might allow future space explorers to cover vast distances, but it might come with a rather deadly side effect...

1 *Shock wave ahead!*
The warp bubble travels through space, carrying the interstellar camper van along with it. Ahead of it, spacetime is so heavily distorted that a shock wave forms (like the wave that forms ahead of the bow of a ship).

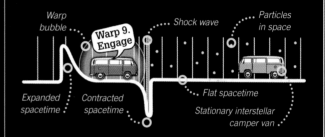

Warp bubble

Warp 9. Engage

Shock wave

Particles in space

Expanded spacetime

Contracted spacetime

Flat spacetime

Stationary interstellar camper van

2 *A full vacuum*
Even though it seems empty, space is full of all sorts of different particles. As the warp bubble moves through space, it picks these up. Some enter the bubble, but others become trapped in the shock wave.

Almost there

Particles build up at shock wave

3 *Kaboooooom!*
These trapped particles pick up huge amounts of energy as they are swept along. When the warp bubble stops, the particles are released in a high-energy beam that destroys everything in its path.

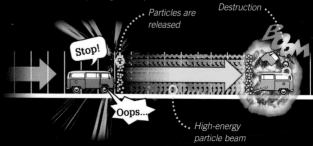

Particles are released

Destruction

BOOM

Stop!

Oops...

High-energy particle beam

SPACE: THE FATAL FRONTIER

SO, AFTER DECADES of careful budgeting, you finally bought your very own interstellar camper van. You are a month into your trip to Mars – to boldly camp on the plains of Amazonis Planitia (with stunning views of Olympus Mons) – when disaster strikes. **Billions of tonnes of radiation**, spewed into space by a colossal solar storm, is bearing down on you.

With no time to move out of its path, the best you can do is **lower your sun visors and hope for the best.** As billions of super-charged particles blast through your body, shattering DNA and obliterating your bone marrow, you have time to regret buying the embroidered seat covers instead of that radiation shield option. Your irradiated, blister-covered corpse is found months later by an itinerant asteroid miner. **You are quite dead.**

If only you had equipped your vehicle with a mini-magnetosphere, plasma radiation shield, things might not have gone so badly. Designed in the early decades of the 21st century, the radiation shield **recreated the magnetic bubble that protects Earth from the worst of the Sun's high-energy hissy fits.** The device consisted of superconducting coils that are supercooled and charged with high-voltage currents to generate a magnetic bubble around the vehicle. This bubble is then filled with a low-density plasma (electrons and protons stripped from hydrogen atoms) that interacts with the magnetic field to create **a cocoon of protective electric currents.** Had you fitted it, the electric field would have been able to absorb or deflect the worst of the solar storm – saving your life.

The device has a long history. First suggested in the 1960s as a means to protect astronauts on long voyages, early designs were dismissed because the electric field would have needed to be too large to be practical. This issue fixed, the shield proved its worth in the 2030s when early lunar and Martian colonists used versions of it to protect their craft and on-surface habitats. **You should have listened to the salesman.**

But how does solar radiation affect the human body, you ask? High-energy particles such as protons and electrons are known as ionizing radiation. They are so energetic that **they can pass clean through the human body**, dumping energy and knocking electrons from atoms (ionizing them). High-energy protons in particular strike molecules in living tissue and break them apart, like teeny tiny bowling balls. Being in the path of a powerful solar event, like a coronal mass ejection (CME), is like **having a neutron bomb go off next to you.** Ionizing radiation wreaks havoc with the structure of DNA that is not easily fixed – creating errors that can lead to cancers. Fast-growing cells like hair follicles, skin, and bone marrow are particularly vulnerable – leading to hair loss, vomiting, diarrhoea, bleeding gums, loss of immune defences, and accelerated aging.

For reasons still not understood by science, crew members wearing red shirts are particularly badly affected and often perish first.

RAISE SHIELDS!

In space, the thin steel shield of the camper van offers no protection from high-energy protons. Even several inches of metal would be pretty useless because protons could cut right through it. In fact, such shielding could be more dangerous than none at all, as protons passing through it can knock neutrons from the shield's atoms – irradiating the occupants with secondary radiation. Several metres of shielding might work, but would be prohibitively heavy. Another solution is required...

EVEN A SMALL SOLAR FLARE EXPLODES WITH THE ENERGY OF MILLIONS OF 100-MEGATON HYDROGEN BOMBS

1 *Electric gas*
The camper van is equipped with a high-voltage machine that tears hydrogen atoms into their constituent protons and electrons. The hot, electrically charged gas (plasma) is then pumped into space around the craft.

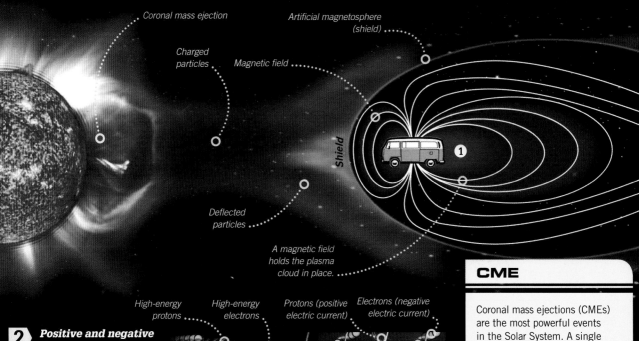

Coronal mass ejection

Charged particles

Magnetic field

Artificial magnetosphere (shield)

Shield

Deflected particles

A magnetic field holds the plasma cloud in place.

High-energy protons

High-energy electrons

Protons (positive electric current)

Electrons (negative electric current)

Shield

Magnetic field lines

2 *Positive and negative*
Electric currents run through the plasma bubble – with negatively charged electrons flowing one way (spiralling around the lines of the magnetic field) and positively charged protons flowing the other.

CME

Coronal mass ejections (CMEs) are the most powerful events in the Solar System. A single CME can throw 10 billion tonnes of charged particles (mostly protons and electrons) into space – covering an area as wide as 48 million km (30 million miles).

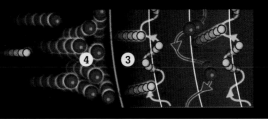

3 *Solar wind*
When the super-charged particles of the solar wind strike the shield, the electrons are captured by the magnetic field and start flowing along the magnetic field lines – boosting the electric current and the shield's effectiveness.

4 *Protected*
The high-energy protons are either stopped completely or deflected by the shield, and flow around the spacecraft – forming a bubble in the solar wind in which the spacecraft is protected.

LEAP SECOND

IT IS ONLY A 💭 THEORY

DEATH RAYS FROM OUTER SPACE

CURIOSITY: SCIENCE'S ❤ HEART

WHY DOES ANYTHING EXIST?

THE STORY OF THE PULSAR

WHAT IS DARK MATTER?

DARK MATTER BUILDS THE UNIVERSE

A WEIRD, ALMOST PERFECT UNIVERSE

IS GLASS A LIQUID?

GRAVITY SLINGSHOT

HELIUM SHORTAGE

THE APPLIANCE OF SCIENCE

WE ARE ALL MADE OF ★ ★ ★ STARS ★ ★

DOING THE BLACK HOLE TWIST

WHY IS 🍎 GRAVITY SO WEAK?

IT IS ONLY A THEORY

SOMETIMES YOU CANNOT HELP THINKING that scientists do not want non-scientists to understand what they are talking about. It is as if they protect their knowledge from the great unwashed by hiding it behind **a minefield of jargon, technical terms, and unpronounceable latinisms** and, as if that were not enough, they have a final line of defence – a smokescreen of linguistic subterfuge where **everyday words become double-agents imbued with confusing and contradictory meanings.**

Of course, this is not the case at all. **Scientists do want to communicate their discoveries** – it is just that sometimes they seem to do so in a slightly different language.

Let us take a look at some words that mean one thing to us but might mean something very different to scientists: **hypothesis, law, and theory.** And then let's see how they work.

Charles Darwin:
In his 1859 book *On the Origin of Species*, Darwin proposed his evolution theory. It stated that all species descended from common ancestors and evolved over time by a process called "natural selection".

EVOLUTION: A PROPER THEORY

In 1835, British naturalist Charles Darwin noted that finches living on different islands of the Galápagos Islands had different beak shapes. Each beak seemed ideally suited to exploit the food source available on each island – finches with large, strong beaks (perfect for nut-cracking) lived on islands with lots of nuts, but, on islands where insects were the main food source, the finches had slender, pointy beaks.

Large, strong beak   Probing beak

Pointed beak   Overbite beak

HYPOTHESIS: »
In his book *On the Origin of Species*, Darwin suggested that a creature's body was sculpted by its environment. Those with features that were best suited to where they lived were more likely to survive, and to pass on those features to their offspring. Over great expanses of time, these small changes could add up to create entirely new species.

Hypothesis

This one is nice and easy and is exactly what you would expect it to be. **A hypothesis is the first rung on the ladder of scientific enquiry. It is an idea, or a best guess, that is formulated to explain observations.** For example, Bob, Sally, and Jake see a curtain flapping around and moving (seemingly) independently of its surroundings. Bob hypothesizes

> IN SCIENCE, A THEORY IS NOT JUST A HUNCH OR A BEST GUESS – IN FACT, IT IS VERY MUCH THE OPPOSITE (NO, NOT A WORST GUESS)

that the movement is being caused by a nearby open window. Sally hypothesizes that there might be somebody hiding behind the curtain whose movements are creating the observed effect. Jake hypothesizes that it must be caused by an invisible, incorporeal spirit.

Bob can test his hypothesis by checking for an open window and then trying to replicate the observation – if it moves when the window is open and stops when it is closed, he can be fairly certain his hypothesis is correct.

Sally can test hers by checking the curtain for a hidden curtain-twitcher – if no one is present, she can formulate an alternative hypothesis. Jake has made an untestable hypothesis – he cannot detect or measure the ghost, so he cannot remove or include its "influence" to test for a result. He might take note of Bob's results and conclude that the open window was the culprit, or he might say the window was only open because the spirit made it so. **Bob and Sally are being scientific – Jake is not.**

Making it law

This where meanings start to get a little muddled. In our society, a law is the pinnacle of a set of rules – a law is the umbrella under which rules reside. But, in science, a law is really only the second rung on the enquiry ladder. **Scientific laws are a description of how something works under specific circumstances.** Taking our example, having successfully tested his open window hypothesis several times, Bob formulates a "law of open windows", which states that an open window will cause a curtain to flap around. Jake will dispute the law because Bob cannot disprove the presence of the curtain-twitching spirit.

Bob's law only describes how the curtain behaves when the window is open; it does not explain what is causing it to behave that way – for that, he has to formulate a theory.

Theory is king

A theory is one of the pinnacles of science and is what scientists like Bob strive to make out of their hypotheses and laws. **A theory usually includes several different hypotheses and laws** – each of which must have withstood

LAW: >>	PREDICTION: >>	EVIDENCE: >>
Biological species change from one kind to another.	For a theory to be successful, it must make testable predictions. Darwin predicted that fossils would be found that would "fill in the gaps" – if one species evolved into another, there must be evidence of the halfway point in its evolution, when it possessed features belonging to both its ancestors and its future descendants.	Just two years after Darwin published *On the Origin of Species*, a fossil was discovered that would become the poster child of evolution: *Archaeopteryx*. Halfway between its dinosaur ancestors and its bird descendants, *Archaeopteryx* shared features belonging to both – just as Darwin predicted.

Archaeopteryx:
Discovered in 1861, *Archaeopteryx* is seen as the "missing link" between dinosaurs and birds.

all attempts to prove them false. **Theories explain observations and laws by providing the mechanism that makes them work.**

Going back to our example, Bob is happy with his "law of open windows" but, as he tests it further, he notices that the rate of the curtain's movement is not constant – sometimes it moves a lot and other times it barely moves at all – so he looks for a mechanism that explains why the curtain moves at all. He develops another hypothesis that suggests that varying air movements outside the open window could account for the variation. He tests this by measuring the air speed outside the window and comparing it to how much the curtain moves. He discovers that there is a connection and develops the "law of air-connected movement", which states that there is a direct correlation between wind speed and curtain movement.

After further testing, Bob discovers that the curtain's movements are also affected by

FOR A THEORY TO BE SUCCESSFUL, IT MUST MAKE PREDICTIONS THAT CAN BE TESTED AND INDEPENDENTLY DUPLICATED

how far open the window is – so he creates a law for this, too. **But all of these laws still lack an explanation of the underlying mechanism that causes the curtain to move.**

After much consideration and many, many more tests, Bob realizes that there must be something within the air itself – something invisible to the naked eye that acts on the material of the curtain, is more energetic in fast-moving air, and is restricted by the aperture of the open window. Based on his evidence, Bob concludes that there must be invisible small bits of matter within the air that, although tiny, can move the curtain when given enough energy. He names the unseen matter after the Latin word for "a small bit" – *particula* – and he calls his new theory the "theory of curtain-moving particles".

Happy with his work, Bob publishes his theory and leaves it to other scientists to search for direct evidence for the existence his "particles".

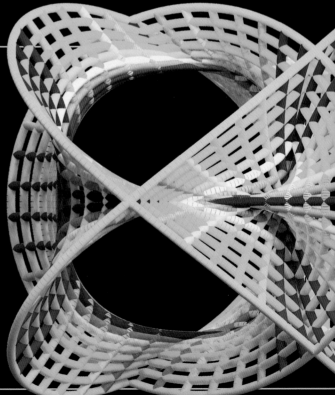

STRING THEORY: THE THEORY THAT ISN'T A THEORY

Two of science's greatest and most successful theories – quantum physics and general relativity – don't work together. Quantum physics fits all the criteria for a theory by predicting and testing effects in the tiny world of particles. Likewise, general relativity works perfectly for predicting and testing how gravity works in the big world of planets, stars, and galaxies. But they are incompatible – gravity can't be explained in the quantum world – and there must be a reason.

Change is good

Some people think that if a theory has to be updated or changed it must be flawed or incorrect. They point out that theories like evolution are always undergoing revisions and are full of gaps in the evidence.

But they misunderstand what a theory is. A theory can be compared to a car. A car has many complex moving parts that perform many different individual tasks, but they all work in harmony to make the car function. Just as a mechanic can upgrade individual parts, add some parts, and take others away without changing the function of the car as a whole, so **scientists can upgrade, replace, and remove hypotheses and laws without changing the overall** truth of the theory. That is the beauty of a scientific theory.

Even seemingly "perfect" theories are subjected to constant tests and observations – if parts are found wanting, they are refined and (if need be) replaced altogether. Science is sometimes accused of self-protectionism and wanting to preserve the status quo to maintain its illusion of infallibility. But **scientists do not test theories to confirm them – they test them to break them.** You will never hear about the scientist who verifies for the 239,000th time Newton's prediction that a feather will fall at the same speed as an anvil in a vacuum. But the chap who finds evidence that gravity is not what we thought it was will become a household name.

GREGOR MENDEL

Born in the modern-day Czech Republic, Gregor Mendel is known as the "father of modern genetics". Farmers had been selectively breeding specific traits into their livestock and crops for generations but the mechanism wasn't understood. Mendel spent eight years experimenting with 10,000 pea plants and concluded that traits are inherited through distinct units called genes. Genes are inherited in pairs – one from each parent. Each gene is either dominant or recessive, with the dominant gene determining the offspring's inherited traits. He worked out a series of laws of heredity, which made predictions that were later tested and replicated by other scientists.

Calabi-Yau manifold:

If they do exist, it is thought that the extra dimensions predicted by string theory would be tightly curled around each other, possibly taking a shape akin to a Calabi-Yau manifold.

HYPOTHESIS: »	LAW: »	PREDICTION:	»	EVIDENCE:
String theory suggests that what we think of as particles of matter are actually vibrating one-dimensional energy strings. Strings vibrating at different frequencies adopt different states – at one frequency one could be an electron, at another frequency it is a photon – resulting in different particles. Likewise, the forces of nature, including gravity, are a manifestation of these vibrations.	Erm... there are none.	This is also where string theory falls flat as a bona fide theory – it makes no (as yet) testable predictions. True, it predicts that there may be many more dimensions than the four we are familiar with – up to 26 dimensions that are curled up so tightly (at a subatomic level) that we are unaware of their existence. The dimensions would exist at such a small scale that it would be impossible to detect them – even with the most powerful of machines.	It also predicts that the familiar subatomic particles should have been created in the Big Bang with an accompanying set of heavyweight cousins, but that these would have disappeared (or decayed) within moments of their creation. Unless these so-called supersymmetry particles are created and detected (before they vanish) in the likes of the Large Hadron Collider, this prediction is also untestable.	There is none. Thus, string theory should not be called a theory.

WHY DOES ANYTHING EXIST?

AT THE DAWN OF EXISTENCE, a mighty war was waged. Two forces faced each other: **matter and antimatter.** Perfect twins separated at birth, but opposite in every way. Neither would be content until the other was annihilated and wiped from the face of existence. Their armies matched each other, particle for particle, and **their mutual destruction should have been assured.**

Yet, against all odds, **matter somehow gained the advantage and emerged victorious.** Our best understanding of the physics of the Big Bang tells us that matter and antimatter were created in equal quantities and, when they made contact in the (far smaller and far denser) baby Universe, all of their combined mass should have been violently transformed into pure energy. Why and how matter survived the encounter is **one of the most profound mysteries in modern science.**

Matter

The current theory is that, although matter and antimatter were created as almost perfect mirror images, there must have been some tiny imbalance, or blemish, that meant that **some were not perfect reflections.** This difference, however tiny, might have been enough to give matter the edge. Scientists have already found a small crack in the mirror, called charge-parity violation, which means that, in some cases, the symmetry of the antimatter reflection becomes broken – **resulting in a particle that is not the perfect opposite of its matter twin.** This "broken symmetry" means that one particle could have an advantage over the other.

This has so far only been witnessed in a tiny number of the particles that are put to the test in the likes of the Large Hadron Collider. But now scientists are pinning their hopes on a less-tested particle – the neutrino.

The neutrino is almost absurdly evasive – it is virtually massless, carries no electric charge, barely interacts with normal matter, and can **spontaneously change its identity,** literally on the fly. It also happens to be one of the most numerous particles in the Universe

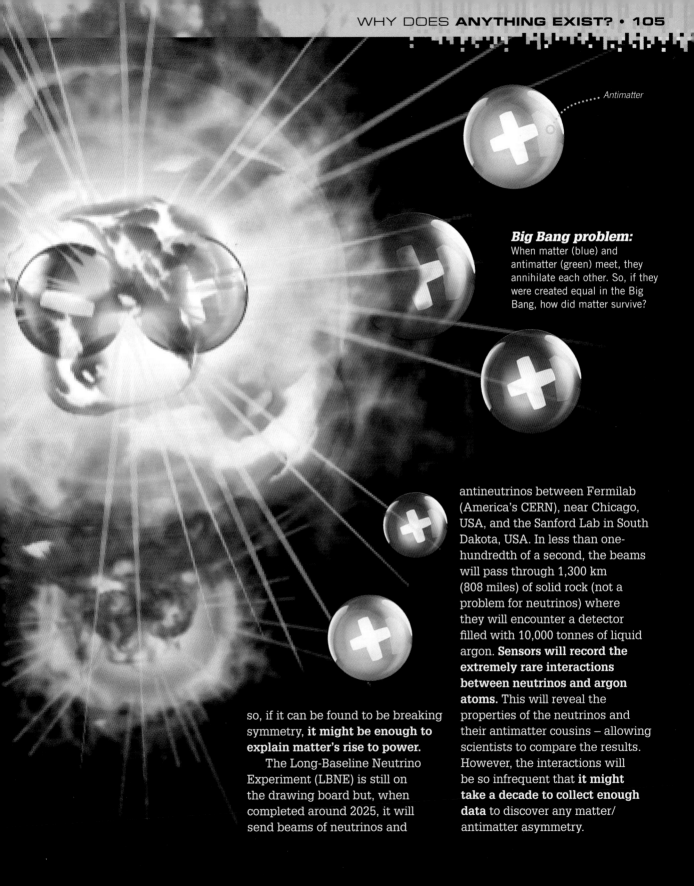

Antimatter

Big Bang problem:
When matter (blue) and antimatter (green) meet, they annihilate each other. So, if they were created equal in the Big Bang, how did matter survive?

antineutrinos between Fermilab (America's CERN), near Chicago, USA, and the Sanford Lab in South Dakota, USA. In less than one-hundredth of a second, the beams will pass through 1,300 km (808 miles) of solid rock (not a problem for neutrinos) where they will encounter a detector filled with 10,000 tonnes of liquid argon. **Sensors will record the extremely rare interactions between neutrinos and argon atoms.** This will reveal the properties of the neutrinos and their antimatter cousins – allowing scientists to compare the results. However, the interactions will be so infrequent that **it might take a decade to collect enough data** to discover any matter/antimatter asymmetry.

so, if it can be found to be breaking symmetry, **it might be enough to explain matter's rise to power.**

The Long-Baseline Neutrino Experiment (LBNE) is still on the drawing board but, when completed around 2025, it will send beams of neutrinos and

BALANCED SYMMETRY

The Big Bang created matter and antimatter together in equal measure. In a perfectly symmetrical universe, where charge and parity are perfectly mirrored, every matter particle would have had an antimatter particle, ensuring their mutual destruction. But that didn't happen...

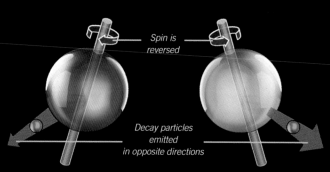

The antimatter version of the negatively charged electron is the positively charged positron

Mass remains the same

Charge is reversed

Negatively charged electron

Positively charged positron

2 **Charge reversal**
At its most superficial level, the antimatter version of a matter particle is one where the mass remains the same, but the electrical charge is reversed. Other properties, such as spin, must also be reversed.

1 **After the bang**
The equal amount of matter and antimatter meant that matter should have been obliterated before anything like stars or planets (or even dust) could have formed – leaving a Universe filled with radiation and nothing else.

Spin is reversed

Decay particles emitted in opposite directions

Left particle

Right particle

3 **Parity reversal**
On the left is a particle that spins to the right, and emits a particle to the left when it decays. Its antimatter partner spins to right, and emits its decay particle to the right. This balancing is known as parity.

RESEARCHING NEUTRINOS

How did matter survive to form the Universe we live in today? The answer may lie with the lowly neutrino. Scientists are building an experiment that will probe the properties of neutrinos and their antimatter cousins as they pass through 1,300 km (808 miles) of solid rock. If they can spot symmetry breaking along the way, it might answer one of science's greatest puzzles.

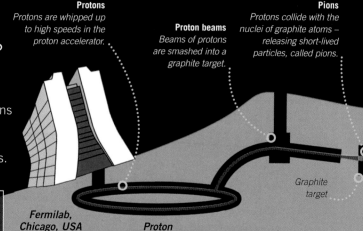

Protons
Protons are whipped up to high speeds in the proton accelerator.

Proton beams
Beams of protons are smashed into a graphite target.

Pions
Protons collide with the nuclei of graphite atoms – releasing short-lived particles, called pions.

Graphite target

Fermilab, Chicago, USA

Proton accelerator

KEY

- Proton
- Muon
- Antineutrino
- Pion
- Neutrino

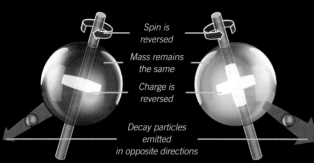

Spin is
reversed

Mass remains
the same

Charge is
reversed

Decay particles
emitted
in opposite directions

**Negatively charged
and left-spinning**

**Positively charged
and right-spinning**

4 *Charge-parity
symmetry*
The perfect antimatter
particle is one that is
an exact mirror image of
its matter equivalent –
having both its charge
and parity reversed.
This is known as charge-
parity symmetry, and is
what we would expect
from the early Universe.

EVERY SECOND, HUNDREDS OF BILLIONS OF NEUTRINOS PASS RIGHT THOUGH YOUR BODY – AS IF YOU AREN'T THERE

VIOLATING SYMMETRY

We now know that symmetry can be broken. Sometimes
an antimatter particle will violate symmetry – perhaps by
emitting its decay particle in the same direction as its
matter partner, or by decaying at a different rate. If
enough violations occurred after the Big Bang, it might
explain why matter survived. By behaving differently from
their antimatter equivalents, it is possible that particles
with broken symmetry just took a little bit longer to decay,
stuck around longer, and so won the day for matter. So far,
these symmetry violations have only been seen to occur
less than 0.1% of the time – not enough to give matter
the upper hand, which is where neutrinos come in.

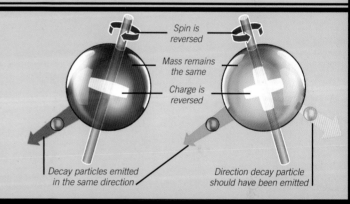

Spin is
reversed

Mass remains
the same

Charge is
reversed

*Decay particles emitted
in the same direction*

*Direction decay particle
should have been emitted*

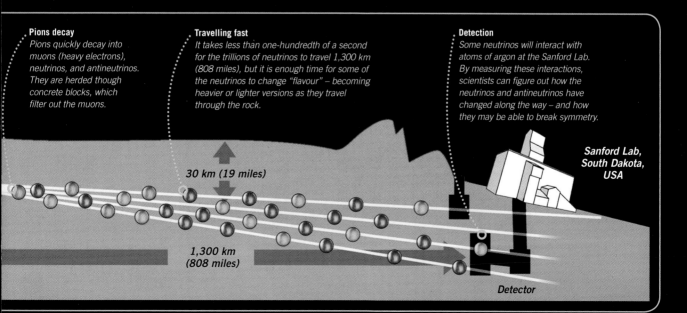

Pions decay
*Pions quickly decay into
muons (heavy electrons),
neutrinos, and antineutrinos.
They are herded though
concrete blocks, which
filter out the muons.*

Travelling fast
*It takes less than one-hundredth of a second
for the trillions of neutrinos to travel 1,300 km
(808 miles), but it is enough time for some of
the neutrinos to change "flavour" – becoming
heavier or lighter versions as they travel
through the rock.*

Detection
*Some neutrinos will interact with
atoms of argon at the Sanford Lab.
By measuring these interactions,
scientists can figure out how the
neutrinos and antineutrinos have
changed along the way – and how
they may be able to break symmetry.*

*Sanford Lab,
South Dakota,
USA*

30 km (19 miles)

**1,300 km
(808 miles)**

Detector

LEAP SECOND

APPROXIMATELY ONCE EVERY YEAR AND A HALF a little extra tick is added to our clocks as the world's official timekeepers decide to **add a "leap second" to the end of the month.** Like a leap year, this leap second is added to bring our clocks back into sync with the rotation of Earth.

The length of a day is determined by Earth's rotation, and one full rotation equals one full day. **But the speed of Earth's rotation is not constant** – ocean tides pulled back and forth by the Moon's gravity, churning molten materials deep in Earth's bowels, earthquakes, and even friction from wind all add up and force the planet to give up a tiny bit of its rotational energy. **In other words, Earth slows down, and our clocks need to compensate for this.**

World time: All sorts of cosmic and terrestrial phenomena conspire to slow the rotation of Earth. Leap seconds are added to compensate for Earth's lagging chronometers.

That's not to say that our planet is not a good timekeeper. Left to its own devices, the day would only lengthen by one millisecond every 100 years, but geological forces, such as earthquakes, can cause the clock to slow. Over millennia, all those tiny increases add up and **in 400 million years or so, a day will be 26 hours long.**

The custodians of humanity's timekeeping are a group called the International Earth Rotation and Reference Systems Service (IERS). These time lords use a global network of radio telescopes called the Very Large Baseline Interferometry (VLBI) network **to measure the speed of Earth's rotation to within a millionth of a second.**

In general, one leap second is added every year or two, but **unusual activities in Earth's core since 1999 have meant that only two leap seconds have been added in this time** (the last was added in 2008).

But why should we care? Would it matter if we let the odd millisecond slide by? True, you and I cannot perceive these variations and, even if we could, it would not really matter. But things like satellite navigation systems have to be able to chart the passage of time so accurately that **these tiny changes do make a difference.**

HOW DO WE MEASURE
TIME SO PRECISELY?

For millennia, mankind was perfectly happy using the Sun to chart the passage of time. But, as technology advanced, so too did our need to track time with ever increasing accuracy. Today's atomic clocks are so precise, they lose less than one second in 300 million years. Here's how they do it:

1 *Jumping electrons*
In an atom, the electrons that surround the nucleus move in orbits that occupy different levels. The electrons can jump between levels.

Electron orbit   Nucleus

Caesium fountain clock
This is the world's most accurate atomic clock – losing less than one second in 300 million years.

2 *Changing levels*
An electron that gains energy moves up a level, and one that loses energy drops down a level. It requires a very specific amount of energy to make the jump. The energy is emitted as electromagnetic radiation at a certain frequency.

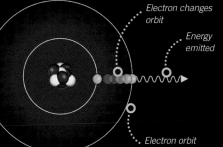

Electron

Electron changes orbit

Energy emitted

Electron orbit

3 *Atomic clock*
An atomic clock uses this wave frequency to chart time (just like an old-fashioned clock uses a pendulum). But, whereas a pendulum only ticks once a second, an atom "ticks" millions of times a second. This means that an atomic clock can chart the passage of time with extreme accuracy.

HOW DO WE KNOW HOW FAST EARTH ROTATES?

How do you judge how fast something is moving when, relative to you, it does not seem to be moving at all? Well, you need to look beyond the object you want to measure and try to judge its speed relative to more distant objects. Imagine an ant sitting on the surface of a slowly spinning playground roundabout. If he were to look at just the roundabout's surface, it would seem that the roundabout is stationary. But if he looked out into the rest of the playground he could judge how fast the roundabout is moving by measuring how long it takes for the swings to pass by, and then (ahem) swing back into view. In the absence of a set of galactic swings, scientists use distant quasars to measure Earth's rotation.

Jets of radiation are belched out by a super-massive black hole at the quasar's centre.

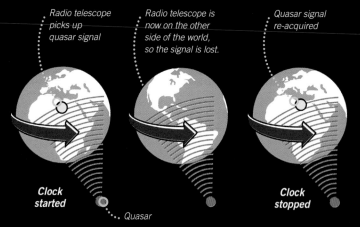

Radio telescope picks up quasar signal

Radio telescope is now on the other side of the world, so the signal is lost.

Quasar signal re-acquired

Clock started

Quasar

Clock stopped

BRIGHT LIGHTS FROM AFAR
Quasars are distant, energetic objects. They are intense sources of X-rays as well as visible light and can shine two trillion times brighter than our Sun (that is 100 times brighter than our galaxy). Most quasars are more than 3 billion light-years away, but they can be even further out than that. Because they are so distant, a quasar's position remains fixed relative to Earth and forms a steady and precise reference point.

MEASURING EARTH'S ROTATION
Scientists use a network of widely spaced radio telescopes to measure Earth's rotation, as well as an ultra-precise atomic clock. Several radio telescopes are pointed towards a distant quasar. When the quasar signal is detected by the telescopes, the time is recorded. As Earth rotates, the telescopes lose the quasar signal. At the end of one full rotation the quasar signal is re-acquired and the time is recorded again, which is then used to give a precise measurement of Earth's rotation.

WHAT WILL HAPPEN IN AN EXTRA LEAP SECOND?

Your body will produce more than **2,500,000 red blood cells**

Mosquitos will infect **8 people** with **malaria**

29,000 bananas will be scoffed

About **40 stars** will end their lives in a **supernova explosion**

Lightning will strike Earth **99,500 times**

255,000 TOILETS will be flushed

A hummingbird will flap its wings up to **200 times**

Hens will lay 2,200 eggs

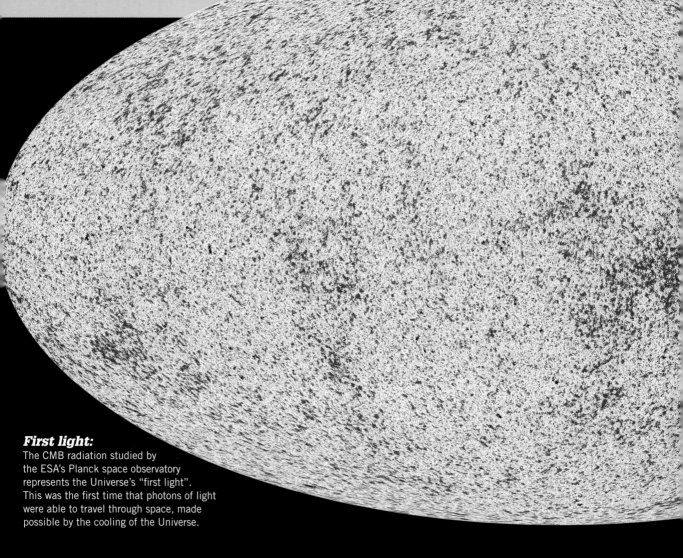

First light:
The CMB radiation studied by the ESA's Planck space observatory represents the Universe's "first light". This was the first time that photons of light were able to travel through space, made possible by the cooling of the Universe.

A WEIRD, ALMOST
PERFECT UNIVERSE

THE COSMIC MICROWAVE BACKGROUND (CMB) dates from about 380,000 years after the Big Bang and represents **the "first light" of the Universe** – released when it had cooled enough to allow photons of light to travel unimpeded through space for the first time.

When the light from the CMB began its journey, **the entire Universe was even hotter than the melting point of iron,** and its energy was emitted as heat – also known as infrared radiation. But, as the Universe expanded, the wavelength of the light was stretched (a bit like how a wavy line drawn on an elastic band becomes stretched when the band is pulled).

The CMB reveals how evenly spread matter and energy were in the early Universe. It also shows how uniform the temperature was: although the colours look dramatic (blue is colder and orange is warmer), they actually represent **temperature differences of less than a hundred millionth of a degree.**

This uniformity of temperature couldn't have been created by a Universe that expanded slowly, so is seen as evidence that **the Universe underwent a period of startlingly rapid, faster-than-light expansion known as cosmic inflation.** By expanding faster than light and information can travel, **space overtook energy's ability to**

react to the change – so, like a rabbit caught in headlights, energy became "fixed" in its pre-inflation state. The only thing missing at this early stage is "dark energy", the mysterious agent thought to be driving the Universe apart at an ever-increasing rate.

CMB: THE FINGERPRINT OF INFLATION

Planck's map of the CMB seems to confirm a key part of Big Bang theory called cosmic inflation. The concept of inflation was added to Big Bang theory in the 1980s to explain the almost perfectly even spread of energy and matter revealed by earlier studies of the CMB.

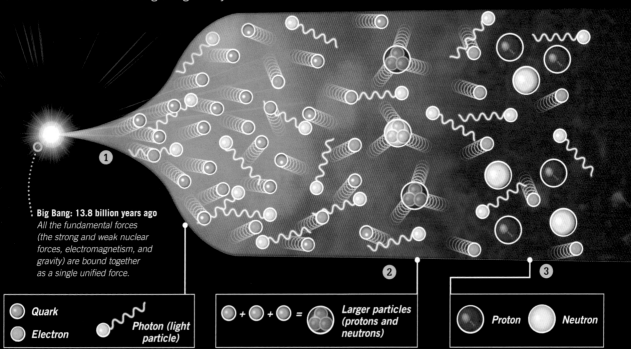

1

Big Bang: 13.8 billion years ago
All the fundamental forces (the strong and weak nuclear forces, electromagnetism, and gravity) are bound together as a single unified force.

2

3

- ⬤ *Quark*
- ⬤ *Electron*
- 〜⬤ *Photon (light particle)*

⬤ + ⬤ + ⬤ = ⬤ *Larger particles (protons and neutrons)*

⬤ *Proton* ⬤ *Neutron*

1 ***Inflation: 0.000000000000000000 000000000000000001 seconds after Big Bang***
Space, time, matter, and energy are all bundled up in an impossibly small, infinitely dense, insanely hot fireball. The Big Bang breaks down the unified force, and powers the exponential inflation of the Universe.

2 ***Fundamental particles: 0.00000000000000000000000000 000001 seconds later***
Energy congeals into matter and the first particles – quarks, electrons, and neutrinos (and their antimatter twins) – are born. These matter opposites collide and annihilate each other, releasing huge numbers of photons.

3 ***Protons and neutrons: 0.000001 seconds later***
As the temperature drops, colliding quarks can join together without being torn apart immediately by all that energy. Quarks combine (via the strong nuclear force) in sets of three to form the first protons and neutrons.

CHANGING WAVELENGTH

When the light from the CMB began its journey 13.81 billion years ago, the Universe was hot, and its energy was emitted as infrared radiation. But, as the Universe expanded, the wavelength of the light was stretched. This stretching has caused its wavelength to move into the microwave part of the electromagnetic spectrum, which is what Planck is designed to detect.

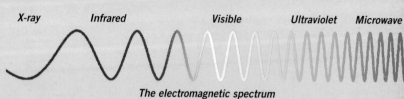

X-ray　　Infrared　　Visible　　Ultraviolet　　Microwave

The electromagnetic spectrum

UNEVEN UNIVERSE

The imperfections in the Universe might be down to something called quantum uncertainty, which tells us that empty space is never truly empty, and therefore never perfectly smooth and regular. Imperfections present at the moment of inflation would have expanded along with the Universe and been imprinted on the Universe from that moment onwards.

CMB IMPERFECTIONS
If matter and energy had been spread perfectly evenly, the Universe we know today would not exist. The temperature fluctuations in the CMB reflect the tiny differences in density and distribution of matter (more matter, more heat). These denser patches had just a little more gravitational pull than their surroundings and accumulated enough matter for the first stars to form.

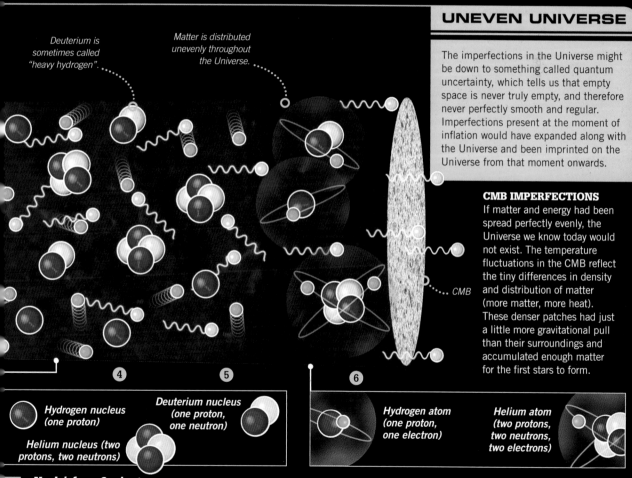

Deuterium is sometimes called "heavy hydrogen".

Matter is distributed unevenly throughout the Universe.

···· CMB

Hydrogen nucleus (one proton)

Deuterium nucleus (one proton, one neutron)

Helium nucleus (two protons, two neutrons)

Hydrogen atom (one proton, one electron)

Helium atom (two protons, two neutrons, two electrons)

4 Nuclei: from 3 minutes until about 377,000 years
When the temperature has dropped to about a billion degrees, colliding protons and neutrons can combine through nuclear fusion to form the nuclei of the simplest chemical elements – hydrogen and helium.

5 Opaque era: from 3 minutes until about 377,000 years
During this era, the Universe is filled with a hot, opaque soup of atomic nuclei and electrons, called plasma. All of the photons created through matter/antimatter annihilations are trapped within the plasma.

6 Stable atoms: 377,000 years later
The Universe cools enough to allow the positively charged atomic nuclei to capture the negatively charged electrons – becoming neutral. With all the nuclei stabilized, photons can travel unimpeded and the

WHAT IS
DARK MATTER?

THE HISTORY OF MANKIND'S RELATIONSHIP WITH THE COSMOS is one of repeated revelations that our place within it is far smaller than we had believed. Once, we thought that Earth was centre of all and the Universe was little more than a **window dressing for the night sky.** Then astronomers revealed our planet to be just one lump of rock travelling around a Sun that is just one star among many hundreds of billions of others in an unremarkable galaxy that is just one among countless billions more.

In a historical heartbeat, **we went from being the kings of a palatial Universe built just for us to an invisible smudge on a speck of matter**, orbiting a mote of incandescent dust, caught in a swirling eddy, lost in the dark ocean of the cosmos.

Dark matter:
This is a computer simulation of the cosmic web of interconnecting filaments thought to underpin the structures of the cosmos. Most of the web is made of invisible dark matter, but about 4 per cent is "normal" matter (the stuff the stars and planets are made of).

FRITZ ZWICKY

Fritz Zwicky was a brilliant Swiss astronomer who, besides dark matter, proposed the existence of supernovae (a name he coined), neutron stars, and galaxy clusters. He also developed some of the earliest jet engines.

Then, just as it was looking as if we had found our (albeit reduced) place in the Universe, **astronomers realized that the way the Universe was behaving did not tally with everything we knew to be in it – something was missing.** So they took measurements and made calculations and concluded that **more than 95 per cent of the matter and energy in the Universe was missing** (well, it was not missing – it was definitely there. We just could not see it).

Humanity's slide down the greasy pole of significance was now complete – a smudge on a speck orbiting a mote of glowing dust in a galaxy afloat in a vast ocean **that makes up just four per cent of the Universe.** Yet, rather than damaging our resolve, each revelation of the vastness of the cosmos has only fuelled our need to understand it better. **Now the hunt is on to find the missing portion of the Universe.**

Why do we think most of the Universe is hiding?

In 1933, a Swiss astrophysicist, Fritz Zwicky, was studying a galaxy cluster (a group of galaxies bound together by gravity). He observed the motions of the galaxies within the cluster and applied Newton's laws to estimate its gravitational mass. But when he came to estimate the amount of visible mass within the cluster (by measuring the light emitted by the stars within it, extrapolating their mass, and adding it all together), **his figures fell drastically short of his first estimate – the visible mass accounted for only a fraction of the cluster's gravitational mass.**

Furthermore, there was not enough visible mass to generate the gravity needed to hold the cluster together (the galaxies should have been flying apart but they were not). **He concluded that there must be something invisible and undetectable making up all that missing mass – dark matter.**

Wimps inherit the Universe

Zwicky's conclusions were initially greeted by the astronomical community with a combination of scepticism and ridicule but, as the years rolled on, evidence for the existence of dark matter began to mount up. **Today, it is almost universally accepted that something beyond our current understanding of physics is at work in the cosmos.** What is not universally accepted though is what form this mysterious something will take. **The term "dark matter" is really just a holding name for whatever it is eventually revealed to be** (rather like the code name an electronics manufacturer might give a gaming console while it is under development).

One of the favourite dark matter candidates is something called a WIMP, or Weakly Interacting Massive Particle. WIMPs are thought to be extremely abundant in the Universe – a billion are estimated to pass through your body every second. As their name would suggest, **WIMPs are expected to possess a lot of mass – possibly the same as an entire atomic nucleus.** Luckily for us, like all wimps, they are really quite shy and barely interact with normal matter – just one out of the many trillions that whizz through you every year might interact with an atomic nucleus within your body.

DARK ENERGY

Dark matter, which makes up about 24 per cent of the Universe, should not be confused with dark energy, which accounts for 71.4 per cent of it. In many ways, dark energy is dark matter's opposite – whereas dark matter holds the Universe together, dark energy is a mysterious force that is fuelling its ever-accelerating expansion.

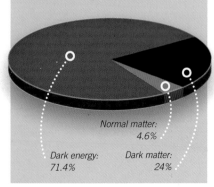

Normal matter:
4.6%

Dark energy:
71.4%

Dark matter:
24%

Their shyness makes WIMPs a perfect dark matter candidate because **something that barely interacts with normal matter would be all but invisible to those of us made of normal matter.** Also, because they are thought to be so massive, they might account for all that mass that we cannot see – they might not physically interact with matter, but their mass means that **matter can feel their gravitational influence** (which is how we know dark matter exists at all).

> IN 2003, NASA OBSERVED A CLOUD OF HOT GAS AROUND A GALAXY CLUSTER THAT WAS ESTIMATED TO CONTAIN A DARK MATTER MASS EQUIVALENT TO MORE THAN A HUNDRED TRILLION SUNS

Unfortunately, the very qualities that make WIMPs ideal dark matter candidates make them **extremely difficult to find** – but, fortunately, not impossible. Once again, the clue is in their name: they are known as "weakly interacting" particles (not "never-interacting"), which means that **every so often they do interact with normal matter** – we just need to catch the moment that a WIMP smashes into an atom of matter. If one does hit an atom, it will nudge the nucleus and give it a dose of energy that the atom doesn't want. **The atom gets rid of the energy by emitting a photon and some electrons that create a flash of light that can be detected by sensors.**

Finding a WIMP

Scientists around the world are now racing to be the first to find this sort of direct evidence of WIMP interactions. To have any hope they need to maximize the chance that a WIMP will collide with an atomic

nucleus, and the best way to do this is to use **something very dense so there are lots of atoms packed together as tightly as possible.**

They also need to build their detectors deep underground to filter out particles of ordinary matter and eliminate the chance that a non-WIMP particle will muddy the results. One such detector, called DarkSide-50, **was completed in early 2014 beneath the Gran Sasso mountains in Italy.** But waiting for a WIMP to come out of its shell in a cave isn't the only method of tracking them down. **Another way is to look to the skies.**

It is thought that all the WIMPs that have ever existed were created a fraction of a second after the Big Bang. Some of them will have decayed into their smaller constituent particles and **some will have been destroyed in high-energy collisions with other WIMPs.** It is hoped detectors launched high into Earth's atmosphere and space will find the particle remnants of the decayed or exploded WIMPs. Scientists working on a dark matter hunter – the Alpha Magnetic Spectrometer (AMS-02), currently installed on the International Space Station – **have hinted that they have found strong evidence supporting the existence of dark matter.** The discovery has sent shock waves through the entire scientific community.

CAN **DARK MATTER** BE FOUND?

You can think of the AMS as being a smaller version of the detectors used at the Large Hadron Collider to analyse particle debris. But, instead of relying on a 27 km (16 mile) ring of magnets to whizz particles up to speed, it uses the world's most powerful particle accelerator: the Universe. AMS uses a series of magnets to bend particles into its detectors in the hope of picking up the electrons and positrons – the antimatter twins of electrons – that are expected to be spat out when WIMPs collide. Here's how it works:

1 *Transition radiation detector (TRD)*
The TRD identifies the sort of particle entering the device. It can tell the difference between an electron and a proton (an electron emits X-rays as it passes through). Without the TRD, the AMS would not be able to tell the difference between a positively charged proton and the electron's antiparticle – the positively charged positron.

2 *Superconducting magnet*
This bends the path of charged particles so they can be identified – a negatively charged particle will bend towards the positive pole, and a positively charged particle will bend towards the negative pole.

3 *Silicon trackers*
These four track the path of particles as they are bent by the magnet.

4 *Electromagnetic calorimeter*
The total energy of the particles is measured here.

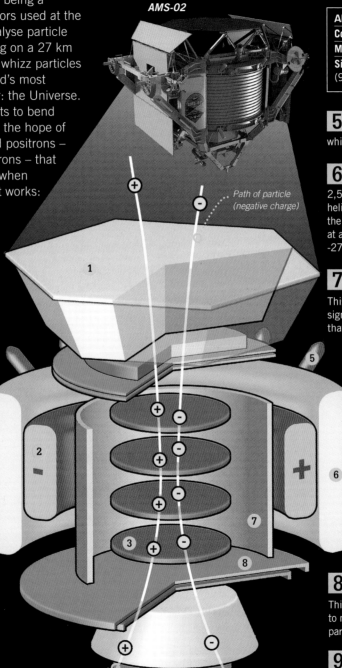

AMS-02

Path of particle (negative charge)

Path of particle (positive charge)

AMS-02 cutaway view

AMS-02
Cost: £1.25bn ($2bn)
Mass: 7,000 kg (15,432 lb)
Size: 3 m × 3 m × 3 m
(9.8 ft × 9.8 ft × 9.8 ft)

5 *Star trackers*
These tell the AMS which way it is pointing.

6 *Helium tank*
The helium tank contains 2,500 litres (5,283 pints) of helium. This is used to keep the super-conducting magnet at a temperature of about -272°C (-457°F).

7 *Anti-coincidence counter*
This tells the AMS to ignore signals from stray particles that enter through the sides.

8 *Time of flight counter*
This acts like a stopwatch to measure each particle's speed.

9 *Ringing imaging Cherenkov detector*
This precisely measures the velocity of each particle.

WHY IS GRAVITY SO WEAK?

ON THE FACE OF IT, GRAVITY would seem to be a pretty impressive force – after all, it is responsible for the formation of planets, stars, and galaxies. Earth and all the other planets of our Solar System are held on an invisible leash and forced by gravity to orbit the Sun. But despite all of this, when compared to the other fundamental forces, **gravity is very puny.**

Gravity is a product of mass – the more massive the object is, the greater its gravitational influence. **Gravity pulls matter towards an object's centre of mass** – planets and stars are round because they are made up of atoms of matter that are **jostling to get as close as possible to a central point.**

What stops all that matter reaching the centre is the electromagnetic force interacting with all those atoms. **Gravity is powerless against the overwhelming strength of electromagnetism.**

For gravity to overcome the electromagnetic force you need a truly massive object – such as a star. Only in the centre of a star is gravity strong enough to force atoms to overcome their electromagnetic repulsion. **But why is gravity so much weaker than the other forces?** No one really knows – and that irritates the hell out of scientists.

Extra dimensions

To explain the mismatch between gravity and the other forces, **physicists have suggested that there may be extra dimensions** beyond the three that we are familiar with – up and down, left and right, forwards and backwards.

For physicists, **an extra dimension is just another direction in space** on top of the three that we humans use to navigate the world. The extra dimensions are hidden from us because of the way we perceive the Universe. String theory **predicts that there are up to 26 dimensions** and that the extra dimensions are hidden from us because they are curled up in really (really, really) small loops.

If that sounds bizarre, imagine an acrobat balancing on a tightrope. In essence, he is occupying a one-dimensional world, in which he can move only backwards and forwards. Now, imagine a flea on the same tightrope. The flea can move backwards and forwards on

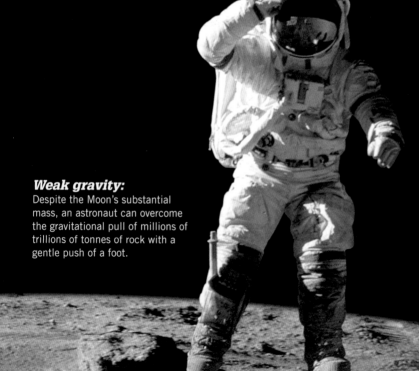

Weak gravity:
Despite the Moon's substantial mass, an astronaut can overcome the gravitational pull of millions of trillions of tonnes of rock with a gentle push of a foot.

the rope, but he can also walk sideways and walk around the rope. The flea is living in a two-dimensional world, but one of these dimensions is a tiny closed loop. **The acrobat can't detect the second dimension, just as we can't detect dimensions beyond the three we move about in.** Also, just as we are trapped within our three-dimensional world, so is everything we use to measure the world around us – such as light and sound. **With nothing interacting with these other dimensions, we have no way of detecting them.**

What does this have to do with gravity?

Physicists have a very effective theoretical framework to describe how the Universe works at the quantum level, called the "standard model". **The theory neatly explains what the fundamental particles do and how they interact with the other fundamental forces.** But, try as they might, physicists just cannot get gravity to fit.

SCIENTISTS ARE STILL LOOKING FOR A "UNIFIED THEORY" OF PHYSICS THAT WILL TIE TOGETHER EINSTEIN'S RELATIVITY AND QUANTUM MECHANICS

Although all the other fundamental forces are trapped in our three-dimensional world, **gravity is thought to be free to travel through these extra dimensions.** As it spreads out through all the extra dimensions it becomes increasingly diluted – making its effect on our three-dimensional world much weaker.

So how can we test this?

Well, according to the standard model, **each of the fundamental forces has a special sort of particle called a force carrier associated with it.** These are like messenger boys that carry instructions to other particles telling them how to be influenced by the force.

It is thought that gravity must also have a force carrier particle, called the "graviton". Sadly, we have never actually seen a graviton, which is where particle colliders like the Large Hadron Collider (LHC) come in.

When the LHC smashes protons together, all sorts of particle gubbins flies out of the energy maelstrom. Given enough energy, **there is a chance that (if it exists) a graviton will be spat out from the collision.** If gravity does permeate all those extra dimensions, there is a chance that **the newly produced graviton will immediately disappear as it escapes into one of them.**

So, our best chance of detecting the extra dimensions (which we can't see) is to find (from among all the other particle mess) a graviton (which may, or may not, exist) disappearing from our plane of existence.

That is the LHC's next big task. **Sounds like a doddle.**

FUNDAMENTAL FORCES

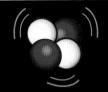

STRONG NUCLEAR FORCE
Power: 1,000,000,000,000, 000,00,000,000,000,000, 000,000,000 times stronger than gravity
Reach: Subatomic
Force carrier: Gluon
This binds matter together. It can't reach very far, but is strong enough to hold protons together within an atom, even though their positive charge is pushing

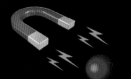

ELECTROMAGNETIC FORCE
Power: 10,000,000,000,000, 000,000,000,000,000,000, 000,000 times stronger than gravity
Reach: Infinite
Force carrier: Photon
Electromagnetism is perhaps the most familiar force as it encompasses everything from magnetism, to light, to the radio waves we communicate with

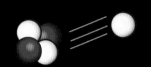

WEAK NUCLEAR FORCE
Power: 100,000,000,000,000, 000,000,000,000,000,000 times stronger than gravity
Reach: Subatomic
Force carrier: W and Z bosons
This is the force responsible for radioactive decay. It allows an atom to change by taking on or losing particles.

GRAVITY
Power: Really weak
Reach: Infinite
Force carrier: Graviton (not yet discovered)

Gravity has a powerful effect on planets and stars, but has almost no influence on matter at the quantum level.

Illuminating the cosmic web: A computer-simulated image of dark matter filaments, which were seen directly for the first time in 2014. These hidden structures form the foundations on which the galaxies where built after the Big Bang.

DARK MATTER
BUILDS THE UNIVERSE

WHEN YOU LOOK UP at the stars that pepper the night sky, or at images of distant galaxies, **you would be forgiven for thinking that they float alone as isolated oases of light in the vast empty ocean of space,** but is the desolate blackness as barren as it appears?

Scientists have believed for some time that **the isolation of the galaxies is actually an illusion.** Instead, they are all connected by a cosmic web of interlinking filaments – huge, invisible highways that carry cold, diffuse gases into the galaxies as fuel for their stellar furnaces.

A legacy of the Big Bang and the tiny energy fluctuations that formed in the Universe's first moments of life, **this cosmic web is thought to have formed the foundations on which the stars and galaxies were built.**

Most of the web (about 84 per cent) is made of invisible dark matter, which **can only be detected indirectly** by measuring the effects its gravity has on the matter (gases, dust, stars, and so on) that we can see. **Luckily, the filaments contain, and are surrounded by, hydrogen gas.** Usually, this gas is too cold and thinly spread to detect with telescopes, but, if it is bombarded with energy, it can be made to glow (like the gases inside a fluorescent light tube).

Now, with the help of a supermassive black hole, these **cosmic filament gases have been observed directly** by astronomers using the 10 m (32.8 ft) wide Keck telescopes in Hawaii, USA.

Illuminated by a nearby quasar (a galaxy with an active black hole that spews out high-energy radiation), called UM 287, the gases are allowing scientists to see the web's filamentary structure for the first time – **confirming the existence of this vestigial remnant of the Big Bang.**

THE WEB

Scientists have believed for some time that space is not nearly as empty as it appears. Instead, all the stars and galaxies are connected by a vast cosmic web of interlinking filaments...

GLOWING QUASAR

In this image, the bright white blob is a quasar. The blue fuzz is glowing hydrogen gas in the surrounding filaments – the first time this has been seen directly. It might not look like much, but this fuzzy blue blob is 2 million light-years wide – in contrast, our Milky Way galaxy is "just" 100,000 light-years in diameter. Although extremely diffuse, the hydrogen gas in this image weighs in at the mass equivalent of a thousand billion Suns.

Supermassive black hole at the centre of a quasar

High-energy radiation

Glowing filament

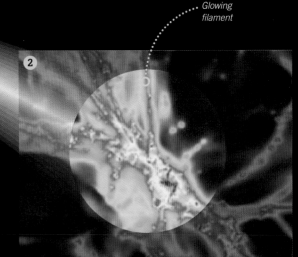

1 **Gas accumulation**
Clumps of matter (mostly hydrogen gas) accumulate where the filaments intersect – forming stars, which collect to make galaxies. One sort of galaxy, called a quasar, has an active supermassive black hole at its centre, which pumps high-energy radiation into the space around it.

2 **Excited gas**
When this radiation bumps into hydrogen gas in nearby filaments, the gas gets all excited and starts to glow – making it visible to our telescopes.

HOW THE WEB WAS SPUN

Following its birth in the Big Bang, the Universe was a roiling soup of blazing plasma. Eventually, this settled down and cooled – forming stable atoms of hydrogen (with some helium and a little lithium). As the Universe expanded, this spread out to become a diffused cloud of gas...

GENERAL RELATIVITY

Albert Einstein's theory of general relativity shows us that gravity is a by-product of mass. Objects with mass (everything from stars and planets to tortoises and particles) bend the fabric of space around them (spacetime) – making "dents" that other, less massive, objects "fall" into. The greater the mass, the deeper the dent and stronger the gravitational pull.

1 Stationary

If this cloud had been spread perfectly evenly, gravity would have acted perfectly evenly on each particle within it. With every particle being pulled (and pulling) the same amount in every direction, they would have remained perfectly stationary.

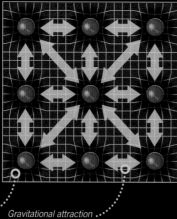

Hydrogen atom •••••   Gravitational attraction •••

2 Density

But matter within the cloud was not spread perfectly evenly. There were tiny imperfections – regions where matter was a little more, or a little less, dense. Regions of higher density exerted slightly more gravitational pull – so particles in less dense regions were drawn towards more dense regions.

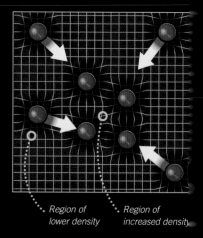

Region of lower density   Region of increased density

3 Gravitational dents

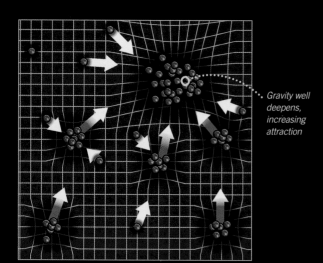

Gravity well deepens, increasing attraction

The more mass that accumulated in one region, the deeper the gravitational "dent" it made in spacetime, and the more mass it attracted.

4 Clouds and filaments

Over millions of years, gas in these regions accumulated into increasingly dense clouds, with connecting filaments. The densest clumps became nurseries for the very first stars and galaxies.

Dense clouds of gas

Gas filaments

5 The true cosmic web

The rapid collapse of gas into the complex web of filaments and dense gas clouds could not have been achieved by the mass of "normal" matter alone. There was too little, spread too evenly over too much space to provide the mass needed to pull everything together so quickly. Luckily, there was lots of dark matter kicking around – with more than enough mass to get the ball rolling. Once these invisible dark matter particles attached to the gas clouds and filaments, the true cosmic web was revealed.

Normal matter "falls" into dark matter clumps called "nodes".

Dark matter filaments act as highways – funnelling matter into the nodes to fuel star formation.

6 The dark force

Dark matter outnumbers normal matter by about six atoms to one. Only dark matter has enough gravitational oomph to form complex structures. Even if normal matter had been able to pull itself together, without the dark matter web holding it in place, it would have been torn apart by the expansion of the Universe.

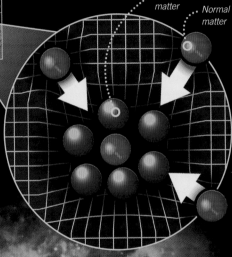

Dark matter

Normal matter

7 Forming galaxies

It was the cosmic web of dark matter that gave normal matter the gravitational foundations it needed to accumulate and build the cities of stars we call galaxies. The interconnecting filaments behave like a transport system – moving matter into the cities to be used to build new stars.

Sharpless 2-106 star-forming region

WE ARE ALL MADE OF STARS

IN THE BEGINNING, THERE WAS THE VOID. The Universe was formless and empty, and darkness was over the surface of the deep. Then there was light and the light was good. **The light was energy and from that energy came matter.** But the matter was simple and disparate, which was not good. Then matter was drawn together and **the first stars illuminated the darkness.** From within the belly of the inferno, simplicity begat complexity and the first heavy elements were born. Hydrogen begat helium. Helium begat carbon and oxygen. Carbon begat magnesium and aluminium and these begat silicon and iron.

Heavy with their elemental progeny, the stars burst forth and spread their seed into the darkness. **From the stars' seed, came forth the Sun and Earth.**

1 *Boom!*
All the matter that will ever exist was created in the Big Bang about 13.8 billion years ago.

BIG BANG!

2 *First particles*
At first it was a roiling soup of energy but, as it cooled, that energy condensed into tiny subatomic particles, which formed the first protons and neutrons. Since hydrogen atoms are made up of a single proton in their nucleus, we now have the first hydrogen nuclei.

Proton (also a hydrogen nucleus)

Subatomic particles

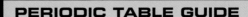

On the land, hydrogen married oxygen and **together they became water.** The elements came together and created complex chemicals and these in turn created amino acids. From the amino acids was brought forth life and **soon the waters were pregnant with living creatures.**

The living creatures were fruitful, increasing in number and filling the waters of the seas, the lands of Earth, and the vaults of the sky.

One of these creatures, called a human being, looked to the heavens and asked, **"where did I come from?"**

BY THE TIME THE UNIVERSE HAD COOLED, IT WAS MADE UP OF ABOUT 75% HYDROGEN AND 25% HELIUM

3 **Helium nuclei**
At this point, the Universe was still very hot and dense – enough to squeeze some of those protons and neutrons together to create the first helium nuclei.

PERIODIC TABLE GUIDE

Elements are arranged in a specific order in the Periodic Table, based on increasing atomic number.

Atomic number (number of protons in the nucleus – the atom's core)

2

He
Helium

4

Chemical symbol

Average mass (including number of protons, neutrons, and electrons in the nucleus)

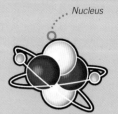

Nucleus

Helium atom (two protons, two neutrons and two electrons)

WHAT ARE WE MADE OF?

An element is a substance that cannot be broken down into any simpler substance by physical or chemical means. This diagram shows the elements that make up the human body by percentage of mass. Only hydrogen was made in the Big Bang – the rest was cooked up in the stars.

65% *Oxygen* O

18.5% *Carbon* C

9.5% *Hydrogen* H

3.2% *Nitrogen* N

1.5% *Calcium* Ca

1% *Phosphorous* P

0.4% potassium, 0.3% sulphur, 0.2% chlorine, 0.2% sodium, 0.1% magnesium, 0.1% iodine, 0.1% iron, and 0.1% everything else Others

Helium nuclei

Helium atom

4 **First atoms**
The Universe cooled down a little more and the electrons – subatomic particles with a negative charge that were made in the Big Bang – were attracted to the positively charged protons, forming the first hydrogen and helium atoms.

Hydrogen atom *Electron*

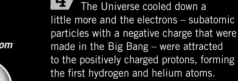

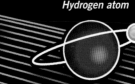

STARS: PARTICLE PRESSURE COOKERS

As the famous American science communicator, Carl Sagan, once said, "we are all made of star stuff". The following shows just how this is true.

(Note: The processes that create some of the heavier elements are vastly more complex than alluded to here.)

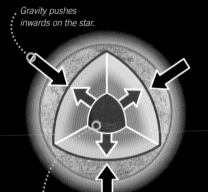

Gravity pushes inwards on the star.

Heat created in the core by nuclear fusion reactions pushes outwards.

Helium is created inside stars when hydrogen atoms fuse.

1 Heat and gravity

In its core, a main-sequence star is fighting a battle between heat – created in the star when hydrogen atoms fuse to create helium – and gravity. Fusion creates energy because the mass of the new helium particle is less than the combined mass of the hydrogen atoms that made it. The leftover mass is released as energy, in the form of light and heat.

Outer layers of the star expand.

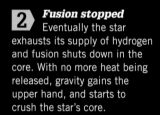

2 Fusion stopped

Eventually the star exhausts its supply of hydrogen and fusion shuts down in the core. With no more heat being released, gravity gains the upper hand, and starts to crush the star's core.

3 Red giant

Even though the core is crushed, the rest of the star expands as its gases cool and the star becomes a red giant.

> **"THE NITROGEN IN OUR DNA, THE CALCIUM IN OUR TEETH, THE IRON IN OUR BLOOD... WE ARE MADE IN THE INTERIORS OF COLLAPSING STARS"**
> CARL SAGAN

4 Fusion restarted

As the core collapses, the pressure and temperature within it rise until fusion can start again – this time with the helium it made earlier. With helium burning away nicely, the collapse is halted and the star settles down into the next stage of its life.

5 New elements

Helium fusion creates two new elements that will turn out very useful when building your body – oxygen and carbon. But before very long (about a million years or so) the star runs out of helium to burn as well. For a star like our Sun, running out of helium is terminal.

Helium starts burning in the star's core.

Gravity continues to try to squeeze the core.

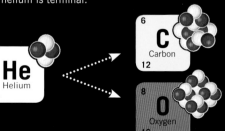

6. Repeat, repeat, repeat

For a star more massive than our Sun, after helium is burnt out, fusion begins anew. The carbon made earlier by the star is fused to create heavier elements such as magnesium, sodium, and aluminium. The processes of fusion, fuel-exhaustion, core collapse, and re-ignition are repeated again and again – each time creating heavier elements until, finally, iron is created.

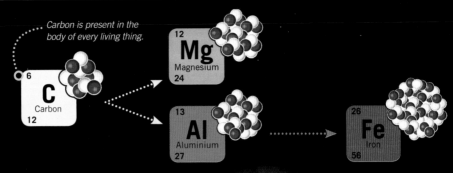

Carbon is present in the body of every living thing.

6 **C** Carbon 12

12 **Mg** Magnesium 24

13 **Al** Aluminium 27

26 **Fe** Iron 56

As the star ages, it cools, expands, and becomes a red giant.

Iron inner core

7. Gravity wins

Even massive stars end their lives here. Iron fusion uses up more energy than it can release and, with no new energy to resist it, gravity is finally victorious and the core collapses for the last time.

The star builds up layers of the elements it has created, wrapped around the core.

Core collapses violently.

8. We are all made of stars

But even this is not the end of the story. The last violent collapse of the core triggers an explosive shock wave that blasts through the star. The shock wave carries so much energy that even iron is fused to create the heaviest naturally occurring elements, such as uranium. The explosion blasts all these elements out into space at thousands of kilometres a second where, one day, they will come together to create new stars and planets and (eventually) you and me.

WHAT ABOUT OUR SUN?

In about five billion years, the Sun will start to run out of hydrogen fuel. It will slowly expand to be about 260 times the size it is now to become a red giant star. This process will swallow the inner rocky planets and, about 7.5 billion years from now, Earth will also be incinerated.

26 **Fe** Iron 52

92 **U** Uranium 238

THE STORY OF THE PULSAR

ALBERT EINSTEIN'S THEORY OF GENERAL RELATIVITY (GR), which describes how gravity is the result of mass, energy, and the curvature of spacetime, has passed every test thrown at it since it was thought up in 1915. But, despite its success, **relativity is not expected to be the last word in gravity.** Although it makes superbly accurate predictions for everyday gravitational objects, relativity **has not been tested in more extreme circumstances.**

You do not get much more extreme than the pair below. The larger object is a fairly unremarkable white dwarf star, but the smaller one, **a newly discovered pulsar,** is an extremely remarkable object indeed.

Imagine an object that could sit quite happily in the centre of New York City, and you could walk around in just a few hours. Now envision that bundled up inside it are **enough atomic nuclei to make two Suns.** Picture its surface burning away at millions of degrees as it shoots high-energy jets of radiation out into space at millions of kilometres per hour. **That is extreme.**

The pulsar, PSR J0348+0432, along with its far less massive companion, is part of a **binary system** in which the two members orbit each other every 2.46 hours. As they plough through space, **they dig gravitational pits in the fabric of spacetime and push up gravity waves,** which spread out into space. According to GR, the binary will lose energy in the process of making those waves and, as such, their orbits will decay, causing them to move closer together.

This prediction has been tested by astronomers using the European Southern Observatory's Extremely Large Telescope in northern Chile. They found that over 12 months of observations, the binary's orbit slowed by eight millionths of a second, which may not sound like a lot, but it is exactly the amount predicted by GR.

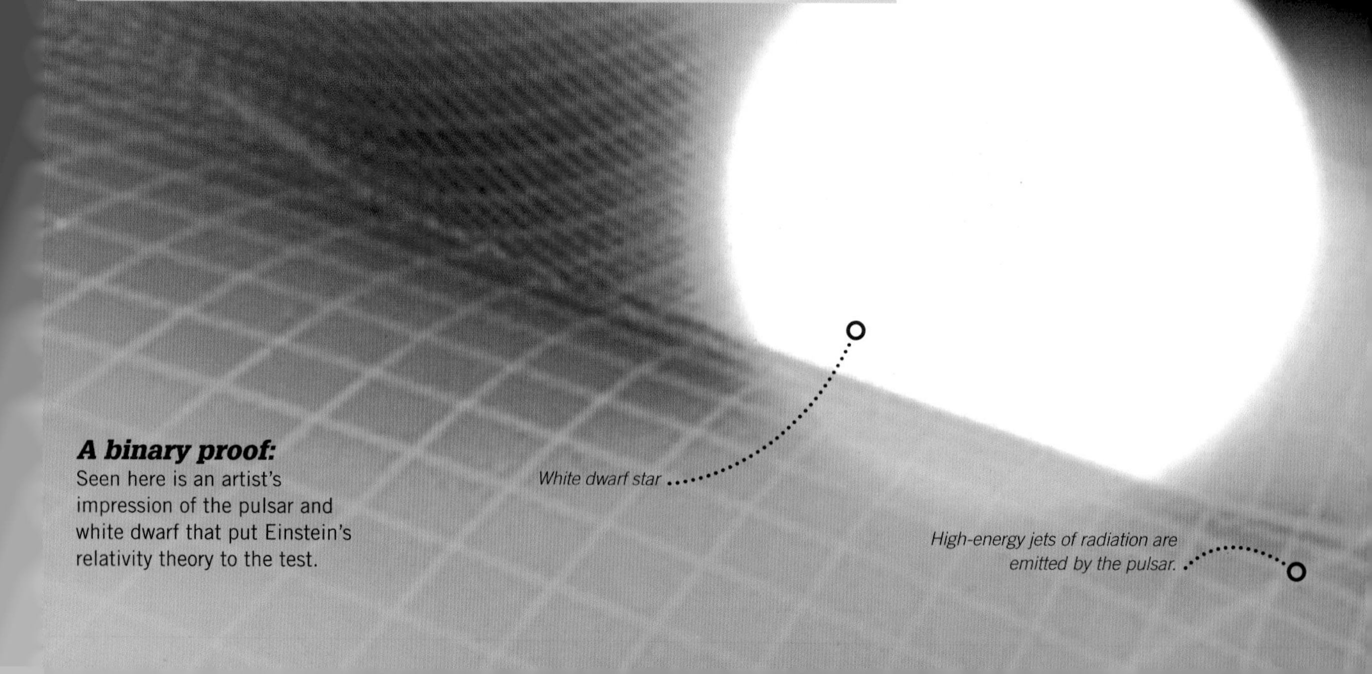

A binary proof:
Seen here is an artist's impression of the pulsar and white dwarf that put Einstein's relativity theory to the test.

White dwarf star ⋯⋯⋯

High-energy jets of radiation are emitted by the pulsar. ⋯⋯⋯○

"STIRRING UP" GRAVITY WAVES

Einstein's theory of relativity tells us that gravity is the result of a massive object distorting the fabric of the Universe (spacetime) around it. The greater the mass, the larger the "dent" made in spacetime and the greater its gravitational influence.

1 Binary system
The pulsar, although tiny, is far more massive than the white dwarf (the spent core of a Sun-size star), so the white dwarf travels around the pulsar.

2 Orbit
The pair orbit together around their shared centre of mass (an imaginary point where their gravity balances out), which, because it is much more massive, is close to the centre of the pulsar.

3 Spacetime waves
As they orbit, they push up waves in spacetime (like a finger stirring the surface of water) that travel out into space.

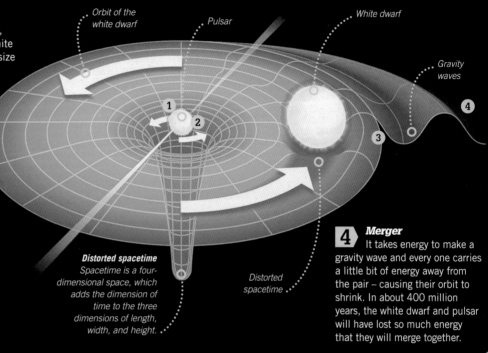

Orbit of the white dwarf

Pulsar

White dwarf

Gravity waves

Distorted spacetime
Spacetime is a four-dimensional space, which adds the dimension of time to the three dimensions of length, width, and height.

Distorted spacetime

4 Merger
It takes energy to make a gravity wave and every one carries a little bit of energy away from the pair – causing their orbit to shrink. In about 400 million years, the white dwarf and pulsar will have lost so much energy that they will merge together.

MASSIVE MASS

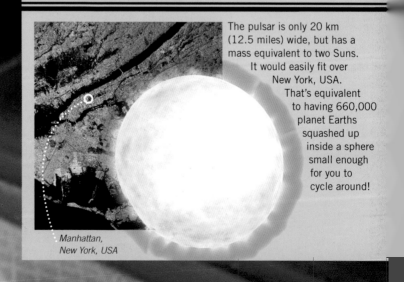

The pulsar is only 20 km (12.5 miles) wide, but has a mass equivalent to two Suns. It would easily fit over New York, USA. That's equivalent to having 660,000 planet Earths squashed up inside a sphere small enough for you to cycle around!

Manhattan, New York, USA

PSR J0348+0432, a pulsar

WHAT IS **A PULSAR?**

A pulsar is a rapidly spinning neutron star with a colossal magnetic field. It emits jets of electromagnetic radiation at rates of up to one thousand pulses per second. If a neutron star exceeds three solar masses, instead of creating a pulsar, its gravity becomes so extreme that it will collapse to become a black hole.

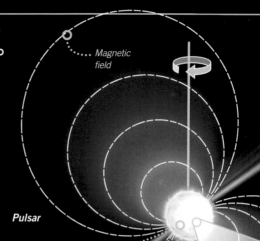

Magnetic field

Pulsar

Rapidly spinning neutron star

Radiation jet

Polar alignment

PULSARS HAVE AN ATMOSPHERE THAT CAN REACH TEMPERATURES OF ABOUT 2 MILLION°C (3.6 MILLION°F)

MASSIVE ATTACK

A pulsar's gravity is so extreme that, if you were to land on one, you would weigh about 7 billion tonnes. But you would not really get the chance to worry about your sudden weight gain because, as you approached the star, you would be stretched into a piece of human spaghetti and fall towards its surface at more than 6.4 million kph (4 million mph). You would then be crushed into a speck of matter smaller than a grain of salt and assimilated into the star's surface.

Atmosphere
A pulsar's super-dense and super-hot atmosphere is just 10 cm (4 in) thick.

Radiation jets
Electrons swirling inside the pulsar emit electromagnetic radiation, which is channelled by the star's magnetic field and emitted as two beams from its poles.

Pulsating jets
The jets are only visible through radio telescopes on Earth when they point directly at us and, as the star spins, the jets spin with it. From Earth, we see the jets as pulses of radiation – hence their name "pulsating stars", or pulsars.

Cross-section of a pulsar

HOW TO BE
MASSIVE, YET TINY

The heart of a pulsar
At its centre is a neutron star. A neutron star is an incredibly dense ball of neutrons created from the collapsed central core of a massive star that ended its life in a supernova explosion.

Outer crust
The extremely thin crust is made of atomic nuclei and electrons (stage 2 in the graphic on the right).

Iron envelope
This is a thin layer of iron atoms.

Outer core
A layer of neutrons that increases in density with depth (stage 3, right).

Inner core
A ball composed of solid neutrons (stage 4, right)

Inner crust
This is made up of crushed atomic nuclei, with electrons flowing through the gaps.

1 **Empty atom**
An atom consists of a nucleus, made up of protons and neutrons, which is orbited by a cloud of tiny electrons. Atoms are mostly empty space (if an atomic nucleus was the size of the one on this page, you would have to scale the electron shell up to the size of a cathedral).

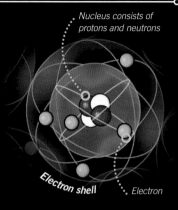
Nucleus consists of protons and neutrons
Electron shell
Electron

2 **Squeezed space**
The negatively charged electrons are kept at this distance from the positively charged nucleus by electromagetic repulsion. But in a neutron star, the gravitational pressure is so extreme that all this empty space is squeezed out of the atom.

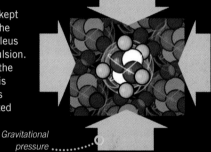
Gravitational pressure

3 **Neutron formed**
Eventually, the pressure becomes so large that electrons are squeezed into the protons – making neutrons (the electron's negative charge cancels the positive charge of the proton, resulting in an electrically neutral neutron).

Electron
Proton
3
Neutron

4 **Neutron star**
The end result is a star made entirely of tightly packed neutrons, which is basically a ball of solid, super-concentrated matter that can pack the mass of an entire mountain range on Earth into a few square centimetres.

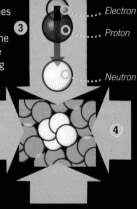
4

IF HUMANITY WERE SQUASHED IN THIS MANNER, WE WOULD BE REDUCED TO THE SIZE OF A SUGAR CUBE

DOING THE BLACK
HOLE TWIST

SO YOU ARE LISTENING TO THE HIT PARADE on the radio and busting some moves on the rug-clad dance floor of your sitting room. A quick glance at your reflection in the glass of your patio doors confirms the **true extent of your awesomeness.** Overcome with exuberance, you perform a spectacular pirouette (or whatever the cool kids call them these days).

Unfortunately, the **surprisingly high frictional coefficient** between your cosy slipper socks and the rug causes it to gather up beneath your feet and, before you know it, you have a rug wrapped around your leg. Everything that was on the rug is likewise **dragged violently inwards**, and half a sitting room's worth of remote controls, discarded socks, and **one confused cat** is hurled towards you... Yup, sometimes spinning sucks.

Supermassive black hole:
Every galaxy is thought to house a supermassive black hole at its centre. Each is powered by a singularity (an almost infinitely dense point in spacetime with the mass of millions or billions of Suns).

But if you think that the cat has it bad, spare a thought for **anything unfortunate enough to be too close to a black hole** when one of these cosmic Travoltas does the twist. Because, instead of messing up a mere woven floor covering, **a black hole drags the very fabric of the Universe along with it** and you can imagine what that does to anything unfortunate enough to be occupying that particular region of the spacetime rug.

Luckily for astronomers wanting to investigate black holes, which by definition are black and therefore virtually invisible, the "twisted spacetime carpet" effect allows them to study black holes indirectly by **looking at the effect they have on the space around them.** Black holes – particularly supermassive ones, which can be found strutting their stuff at the centre of most

galaxies – interest astronomers as they hold clues to how galaxies evolved in the first place.

Astronomers have recently made a supermassive breakthrough by finding a new way **to measure how fast black holes spin.** Armed with the European Space Agency's XMM-

> ## A BLACK HOLE DRAGS THE VERY FABRIC OF THE UNIVERSE ALONG WITH IT

Newton satellite, they took a look at a supermassive black hole with the mass of 10 million Suns that lies at the heart of a galaxy 500 million light-years away.

Like many black holes, it is **surrounded by a spinning disc of gas and dust** that sits like a picnic, spread out across the spacetime rug

waiting to be devoured. By looking at the picnic (also **known as an accretion disc**) the team could determine how far the inner edge of the disc was from the black hole.

This distance tells astronomers how fast the black hole is spinning because **material in the disc is drawn closer as the black hole's spin increases.** The disc was found to be far from the edge of the black hole, which means that, for the moment at least (bearing in mind it is 500 million light-years away so they are studying it as it appeared half a billion years ago), it is spinning at the relatively slow speed of "only" half the speed of light.

But who cares how fast a black hole spins? Well, if you thought a pair of rubber-soled socks can make a mess of a carpet, just take a look at what a spinning black hole can do...

HOW BLACK HOLES WORK...

A black hole is a collapsed remnant of a dead star's core. It is in a star's nature to spin and, when it dies, this spin is transferred to its core. As it collapses under the weight of its own gravity, the core's spin accelerates. By the time it has become a black hole it can be spinning at almost the speed of light.

This is a supermassive BLACK HOLE...

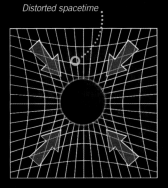

Distorted spacetime

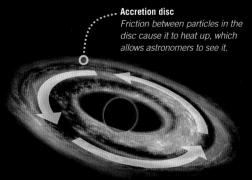

Accretion disc
Friction between particles in the disc cause it to heat up, which allows astronomers to see it.

1 Heart of the matter
At a black hole's heart, locked away from time and space, there is a teeny tiny singularity – a speck smaller than an atom that contains the mass of millions of Suns.

2 So much mass
With all that mass, the black hole bends the fabric of the Universe, making a gravitational dent so deep that not even light can escape it.

3 Un-compact disc
The accretion disc is a swirling disc of gas and dust that builds up around the black hole. If the black hole is just chilling out, the orbital momentum of the material in the disc stops it from falling in.

4 Spin spin spin

As the black hole spins, it drags the fabric of the Universe (spacetime) around with it. Space itself gets all twisted up around it, like a sheet caught in a spinning drill bit – a process known as frame-dragging.

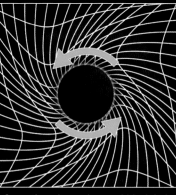

Spacetime dragged around black hole

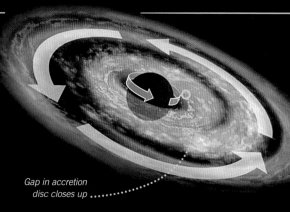

Gap in accretion disc closes up ....

5 A real drag, man

The faster the black hole spins, the more the material in the disc is dragged closer. By measuring the gap between the black hole and the accretion disc, astronomers can figure out how fast the black hole is spinning.

Event horizon
At the event horizon, the accretion disc material is superheated and torn apart by friction, creating a kind of particle blender.

Accretion disc
Material in the disc can be accelerated to insane speeds and thrown past the event horizon and into the black hole, churning up spacetime in the process. ....

6 No escape

The event horizon is the point where the black hole's gravity becomes so extreme that not even light can escape. Beyond it, spacetime is falling into the black hole faster than the speed of light. This is why light cannot escape – space is flowing inwards faster than light can move outwards.

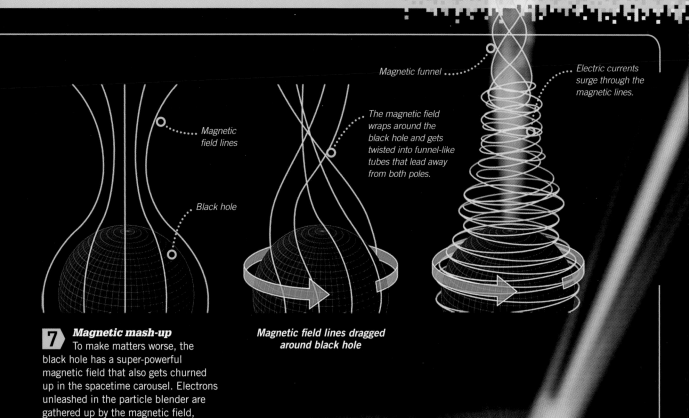

Magnetic funnel . . . .

Electric currents surge through the magnetic lines.

The magnetic field wraps around the black hole and gets twisted into funnel-like tubes that lead away from both poles.

Magnetic field lines

Black hole

Magnetic field lines dragged around black hole

7 Magnetic mash-up
To make matters worse, the black hole has a super-powerful magnetic field that also gets churned up in the spacetime carousel. Electrons unleashed in the particle blender are gathered up by the magnetic field, creating powerful electric currents that surge through the magnetic field lines.

A BLACK HOLE CAN EMIT MORE ENERGY THAN A HUNDRED BILLION SUNS

Particle jet . . . .

8 Radiation station
Particles pulled apart by the spacetime blender are sucked up by the funnel, accelerated by the electric currents and blasted out into space as focused beams of charged particles and radiation.

HELIUM
SHORTAGE

THE UNIVERSE WAS BORN AS A ROILING SOUP OF ENERGY about 13.8 billion years ago in the event known as the "Big Bang". When things had cooled down a bit, the energy condensed into the first particles, and for the first few hundred million years or so, the entire Universe was a vast cloud of hydrogen and helium gas.

Today, despite the best efforts of the stars to convert them into heavier elements, **hydrogen and helium still dominate the mass of the cosmos.**

THE SUN CONVERTS 655 MILLION TONNES OF HYDROGEN

Helium – the second lightest and second most common element behind hydrogen – still accounts for about 24 per cent of the mass of the observable Universe (almost a quarter of everything everywhere). **Yet, here on Earth, it is incredibly rare – making up just 0.00052 per cent of our atmosphere – and our supplies are running out.** By some estimates, Earth's helium reserves could be exhausted within just 30–50 years.

Now, we all know the hilarity that ensues when a party-balloon-toting joker uses the gas to perform dubious Mickey Mouse impressions, but, believe it or not, helium has a serious side.

Capable of remaining liquid at temperatures as low as -269°C (-452°F), it is the most effective refrigerant in the world. It cools the superconducting magnets that power the likes of the Large Hadron Collider and the magnetic imaging scanners used by doctors to peer inside the human body. It is used to make the semiconductors found in virtually every electronic device you take for granted—and the fibre optic cables essential for high-speed Internet, communications, and TV need the cooling power of helium to prevent signal-destroying bubbles from forming during their manufacture. **So running out of helium is certain to cause us a few problems.** In fact, in 2012, medical scanners at some UK universities were all affected by helium shortages. But **why is there so little of it on Earth when the cosmos is literally swimming in the stuff?**

Its main problem is its inert chemical nature. Chemical elements form bonds by sharing electrons, which they do to achieve a sort of Zen-like state of electromagnetic balance. A helium atom has two negatively charged electrons in its shell, and it really doesn't need any

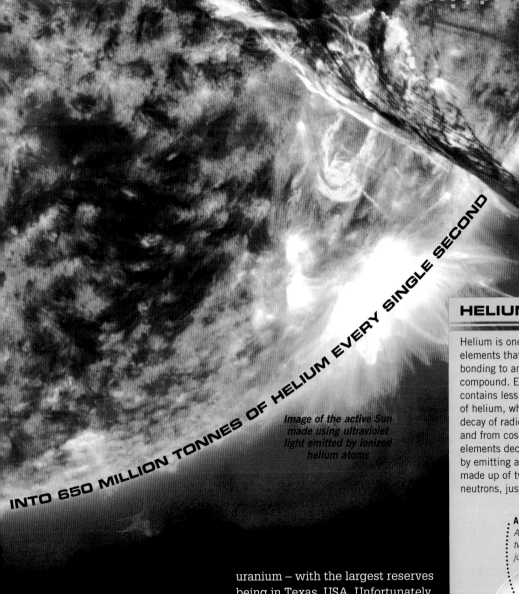

INTO 650 MILLION TONNES OF HELIUM EVERY SINGLE SECOND

Image of the active Sun made using ultraviolet light emitted by ionized helium atoms

HELIUM ATOM

Helium is one of only two natural elements that has never been observed bonding to another to create a compound. Earth's atmosphere contains less than five parts per million of helium, which is resupplied from the decay of radioactive elements on Earth and from cosmic rays. Radioactive elements decay into lighter elements by emitting alpha particles – which are made up of two protons and two neutrons, just like a helium nucleus.

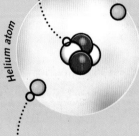

Alpha particle
A helium nucleus contains two protons and two neutrons, just like an alpha particle.

Helium atom

Catching electrons
If an alpha particle can attract two electrons from somewhere, a helium atom is formed.

more (it can be considered to be "full") – which means **it has no incentive to bond with other elements.** Also, because it is so light, any that finds its way into the atmosphere eventually just drifts off into space – a quality that also makes it devilishly difficult to store.

Virtually all our helium comes from underground – as a by-product of the decay of naturally occurring radioactive elements like

uranium – with the largest reserves being in Texas, USA. Unfortunately, the USA (which holds 80 per cent of the world's reserves) has been selling off helium cheaply since 1998 – leading to frivolous usage and wastage. You can get it from the decay of tritium (a radioactive isotope of hydrogen), but the USA stopped making that in 1988.

Luckily, if we do run out of helium on Earth, **there are plenty of other places in the Solar System that have loads of it.**

HOW THE SUN
STOLE OUR HELIUM

Helium is the second most plentiful element in the Universe, and there's no shortage of it in the Solar System either. The Sun contains about 557 million billion billion tonnes of helium and it makes hundreds of millions of tonnes more every second. So why is there a shortage here on Earth?

1 *Gassy origins*
Our Solar System began life in a giant cloud of gas and dust known as a star-forming nebula.

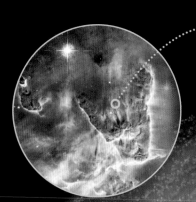

Star-forming nebula

Protostar Sun

2 *Hungry Sun*
Gravity caused part of the cloud to collapse into a swirling disc of gas. At the centre, a protostar that became our Sun began to grow. The baby Sun was hungry for power and, as it grew, it absorbed 99.86 per cent of the mass of the disc – leaving just 0.14 per cent for the planets to fight over.

Disc of gas and dust

3 *Solar wind*
But it was not a fair fight because, as the Sun grew in strength, it started blasting high-energy radiation into the disc. This solar wind pushed all the lighter elements away – banishing all the hydrogen and helium to the outer regions of the disc.

Hydrogen and helium move to the outer part of the disc.

Snow line

4 *Snow line*
A dividing line was created, called the "snow line". Inside the snow line, only the elements heavy enough to resist the solar wind remained – creating a "hot", rocky wasteland.

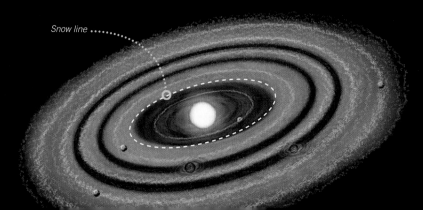

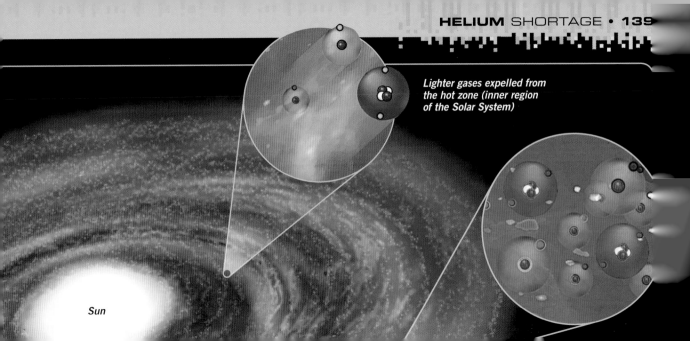

Lighter gases expelled from the hot zone (inner region of the Solar System)

Sun

DEATH RAYS
FROM OUTER SPACE

BACK IN 1901, SOME ENGLISH SCIENTISTS noticed a puzzling thing while experimenting with the radioactive element radium **(radioactivity itself had only been discovered five years earlier).** They were measuring its radioactivity using a gold leaf electroscope, which used an electric field to hold two strips of gold leaf apart.

When "radium rays" entered the device, they ionized the air around the gold leaf, which allowed electrical charge to escape – the more radiation present, the less charge the gold leaf held and the closer the two strips would move together.

The scientists noticed that, even when the electroscope was removed from the radium, it still lost electrical charge. **Somehow radiation was coming from somewhere else.** Nor was this an isolated event. Laboratories all over the world were reporting the same phenomenon.

WHERE DO **COSMIC RAYS COME FROM?**

Because the particles that make up cosmic radiation are electrically charged, they are deflected by magnetic fields. So, as they journey through the galaxy, their paths become scrambled – making it virtually impossible to trace their origins. But that is not to say we do not have some idea…

SUPERNOVAE

It's possible that, when a massive star explodes, the expanding shock wave accelerates the charged particles that are emitted. Trapped inside the remnant's magnetic field, the particles bounce around until they reach near-light speed and escape as cosmic radiation.

Crab Nebula (supernova remnant)

BLACK HOLES

It is thought that the highest-energy cosmic rays are created by supermassive black holes. When matter, or even a star, is devoured by a black hole, it can be spewed out in colossal jets at near-to-light speeds.

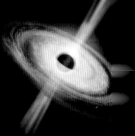

Dangerous combination: Cosmic radiation is made up of about 89 per cent hydrogen nuclei, about 10 per cent helium nuclei, and about 1 per cent really tiny stuff, such as electrons, and the nuclei of heavier elements.

DESTROYER!

Cosmic radiation is made up of extremely energetic particles travelling at close to the speed of light. Some of these can possess the same energy as a tennis ball travelling at 160 kph (100 mph). All this energy can cause a lot of problems…

• Cosmic rays can damage the electronic circuits of spacecraft by causing computer memory bits to "flip" or microcircuits to fail. It can also harm astronauts' DNA. NASA estimates that every week spent in space shortens their life expectancy by a day.

• Some scientists think that many of Earth's extinction events could have been partly caused by cosmic radiation. High levels may have caused genetic mutations – making organisms less able to cope with environmental changes.

STARS

The Sun (and other stars) can produce cosmic rays. Atomic nuclei and electrons can be accelerated by shock waves travelling through the Sun's atmosphere (corona) and by magnetic energy released in solar flares.

GRAVITY SLINGSHOT

WE ARE USED TO THINKING OF SPACEFLIGHT as a struggle against gravity. After all, it takes vast, towering rockets filled with hundreds of tonnes of explosive liquids and gases just to give light aircraft-sized vehicles enough thrust to break free of the bonds of Earth's gravity.

Even if you are lucky enough to make it into space, there are still endless gravitational hurdles to overcome. Contrary to what Sir Isaac Newton believed, gravity is not caused by two massive objects pulling on one another. Instead, **gravity is a by-product of the dents and distortions made by massive objects in the fabric of the Universe.** A truly massive object, like a planet, makes a pretty big dent and, when a less massive object, like a spacecraft, strays too close, it finds itself "falling" into that dent – it might look as if the spacecraft is being "pulled" towards the planet, but really it is "falling" towards it.

The Solar System is littered with these gravitational pitfalls – **a satellite falls towards Earth, Earth falls towards the Sun and, in turn, the Sun falls towards the centre of the Milky Way.** The only way to stop this fall from becoming a direct plunge is to move through space fast enough to ensure your momentum keeps you aloft.

You can think of the Sun's gravity as being a little like a wine glass. If you drop an olive into the glass, it will fall straight to the bottom, but, if you spin the glass, you can give the olive enough momentum to roll around the sides without falling in (like a planet orbiting the Sun). Decrease the momentum and its orbit will fall closer; increase it and its orbit moves further away. **If you continue to increase the speed, eventually the olive will move so fast that it will achieve "escape velocity" and fly from the glass**.

A spacecraft leaving Earth has been given enough momentum to escape Earth's gravity wine glass, but, if it wants to travel into deep space, **it has to find enough momentum to escape the Sun's gravitational dent, too.** Using rockets is not practical because they would need so much heavy fuel that it would be prohibitively expensive just to leave Earth – so **scientists came up with a clever trick called a "gravity assist" manoeuvre.**

Also known as the "slingshot" manoeuvre, the technique was first used successfully 40 years ago by NASA's *Mariner 10* Mercury probe. Instead of struggling against the gravitational pull of the planets, during a gravity assist, a spacecraft uses a planet's (or a series of planets') gravity to give it a speed boost. By falling towards a planet that is falling towards the Sun, **a spacecraft can "steal" enough momentum to travel against the Sun's gravitational pull.** So you could say that spaceflight is not flying at all: it is just falling, with style!

MARINER 10 WAS THE FIRST CRAFT TO USE SOLAR WIND AS A MEANS OF PROPULSION AFTER ITS THRUSTERS RAN LOW ON FUEL

YURI KONDRATYUK

The method of using a celestial body's gravity to accelerate and decelerate was suggested by Ukrainian scientist and engineer Yuri Kondratyuk in his elaborately titled paper "To whoever will read this paper in order to build an interplanetary rocket". The paper was dated 1918–19, but published in 1938, so it may have been preceded by that of a German-born Russian scientist, Friedrich Zander, who made a similar suggestion in 1925. The idea was refined by NASA scientist Michael Minovitch for *Mariner 10*'s trip to Venus and Mercury in 1973.

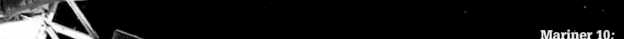

Mariner 10:
NASA's *Mariner 10* probe became the first spacecraft to use the gravitational slingshot effect to reach another planet. On 5 February 1974, it passed by Venus and used the planet's gravity to send it on its way towards Mercury.

USING GRAVITY
TO GO FASTER

Gravitational assist manoeuvres, or slingshots, are an essential part of space exploration. By stealing gravitational energy from a planet, a spacecraft can reach much higher speeds, using less fuel, than would be possible using rockets alone.

SPEED UP, SLOW DOWN

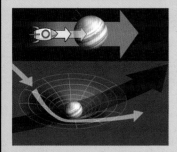

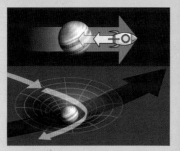

Speeding up: If the spacecraft approaches from behind a planet, it gets a gravitational tow and "slingshots" around the back – gaining speed.

Slowing down: Assists can also be used to slow a spacecraft. By approaching the planet from the front, the craft uses the planet's gravity like an anchor to slow itself.

CASSINI'S TRAJECTORY

NASA's Saturn-exploring *Cassini-Huygens* spacecraft weighed in at 5.6 tonnes – too heavy for even the most powerful rockets to provide enough thrust to counter the Sun's gravitational pull. To provide the extra boost it needed, *Cassini*'s 6.7-year route took it twice past Venus and once past Earth and Jupiter. With each planetary fly-by, *Cassini* stole a little momentum to give it the boost it needed to break away from the Sun's gravity and reach Jupiter.

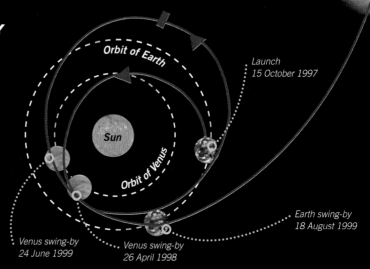

Cassini

Saturn arrival
1 July 2004

Orbit of Jupiter

Jupiter swing-by
30 December 2000

Orbit of Earth

Launch
15 October 1997

Sun

Orbit of Venus

Earth swing-by
18 August 1999

Venus swing-by
24 June 1999

Venus swing-by
26 April 1998

> ## AN ADVANCED INTERSTELLAR CRAFT MAY ONE DAY USE THE GRAVITY OF TWO NEUTRON STARS TO ACCELERATE TO THE ASTONISHING SPEED OF 291 MILLION KPH (181 MILLION MPH)

SUPER SLINGSHOT

Using gravity to get a speed boost sounds straightforward: a spacecraft is accelerated by a planet's gravity and leaves moving faster than it arrived. But, from the planet's viewpoint, the craft also has to leave that planet's gravity and, in the process, it appears to be slowed down to its original pre-approach speed. So where is the boost coming from? It all becomes clear from the Sun's point of view...

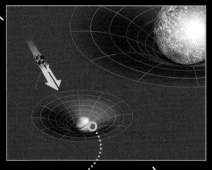

Jupiter

1 **Coming in**
From the Sun's point of view, the craft approaches the planet (in this case, Jupiter) at its normal speed.

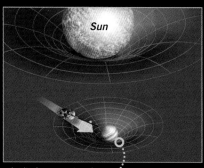

Sun

Jupiter's momentum is added to the craft's.

2 **Gravity well**
The craft accelerates as it falls into Jupiter's gravity well. Jupiter's gravity is now acting on the craft, pulling it with the gas giant as it orbits the Sun.

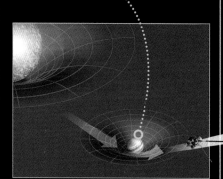

Jupiter has been slowed down a tiny amount.

3 **Gained momentum**
The question of where the boost comes from is all about perspective. It is like a tennis ball being hit at a moving train: if you were on the train, the ball would appear to bounce back at the speed it was thrown. If you were beside the track, you would see that the ball added the train's speed to its own as it bounced off.

MISSION

The *Cassini-Huygens* mission was launched on October 15, 1997 and arrived at Saturn in July 2004. The *Huygens* probe, built by the European Space Agency (ESA), entered the atmosphere of Saturn's largest moon, Titan, and parachuted to the surface - taking detailed readings of the atmosphere and capturing the first images of the moon's surface. *Cassini*'s mission is to explore the Saturnian system – focusing on Saturn's rings and icy moons. The craft has made many orbits of Saturn and fly-bys of Titan, and is expected to continue operating and making discoveries until 2017 at the earliest.

BLACK HOLE NO-GO

You might think that if the gravity of a planet or star can be used to boost a spacecraft, a black hole that contains the mass of millions of stars would make the ultimate slingshot machine. Unfortunately, black holes are so massive and spacetime becomes so heavily distorted around them, that it would take more energy to escape the pull of a black hole than could ever be added by its motion.

IS GLASS A LIQUID?

GLASS IS ALL AROUND US.
It is the window we gaze out of longingly on a warm summer day, and it is the computer screen behind which lurks the work that prevents us from going outside. It is the modern office block in which we toil and it is the empty bottle of wine we discard at the end of a long week. **In the modern world, glass is ubiquitous, essential, and deeply mysterious.**

The mystery of glass started when people looked at centuries-old windows and observed that **their panes seemed to be thicker at the bottom.** This **led to the speculation that glass is a liquid** that in short timescales seems to be solid but that, over centuries, acts like warm toffee – flowing downwards into the encouraging arms of mother gravity.

Shattered glass:
It's generally seen as common sense that liquids can't shatter – if they could, it would make diving into a swimming pool a rather hazardous pastime. So it stands to reason that glass, which definitely shatters, can't be a liquid... or does it?

This, of course, was not true. If glass was so fluid, older glass objects – like millennia-old drinking vessels – would have long ago melted into glassy puddles. The thickening observed in old windows is **simply a result of the manufacturing processes of the time,** which involved spinning molten glass out into sheets, which created glass with thick edges.

But the mystery does not end there. When you look at glass on a molecular level, **it does indeed appear to be more liquid than solid.** In a solid, the molecules are arranged in neat and rigid shapes, but in glass, the molecules are arranged in an almost random jumble – like a liquid. In fact, **structurally there is almost no difference between a liquid and glass.** Glass is made by cooling a liquid below its freezing point. As the temperature drops, the molecules become more sluggish and the liquid becomes more viscous until they become almost motionless and the glass is formed.

But, unlike a solid whose molecules stop moving, the molecules in glass never really stop. It is this lack of transition between its phases that has led many physicists to argue that glass is neither a liquid nor solid, but is instead in a sort of in-between state known as an "amorphous solid".

IF GLASS IS A LIQUID, IT WOULD TAKE 100,000,000,000,000,000, 000,000,000,000,000 YEARS FOR A PANE TO SAG SLIGHTLY

WHAT MAKES
GLASS DIFFERENT?

Glass's molecular structure sits somewhere between a liquid and a solid. Its molecules are jumbled randomly, similar to those in a liquid. But they move a lot slower, to the point where they are almost not moving at all, in a similar state to a solid.

In a solid, the molecules adopt a rigid crystalline structure.

But glass, although seemingly solid, has the random jumbled structure of a liquid.

WHAT'S THE MATTER?

Unlike most substances, which change from their liquid state to their solid state at a set temperature – water solidifies into ice at 0°C (32°F) – the temperature at which glass forms depends on how quickly it is cooled from its molten state. The slower you cool it, the lower the temperature at which it changes into glass and the more dense that glass will be – cool it quickly and it changes into a solid while it is still much hotter. Whether you believe that that makes it a solid or a liquid, or both, glass isn't the only weird state of matter...

SUPERFLUIDS

A superfluid is a phase of matter that occurs when helium is supercooled to temperatures close to absolute zero, which is about -273.15°C (-459.67°F). At this temperature, its atoms exhibit weird quantum effects that give it infinite heat conductivity and zero friction. A superfluid isn't affected by surface tension and can seemingly defy gravity by "climbing" out of its container.

DARK MATTER

This is a mysterious form of matter that makes up 83 per cent of the matter mass of the Universe. Because it doesn't interact with the electromagnetic force, it doesn't emit or absorb radiation – meaning it is invisible and (so far) impossible to detect directly. It has been detected indirectly through its gravitational effects on objects we can see (like stars and galaxies).

DEGENERATIVE MATTER

Under "normal" conditions, atoms are made up of almost entirely empty space (the space between an atomic nucleus and its electron cloud can be likened to a mosquito inside a cathedral). When matter is subjected to intense pressure, all that empty space is squeezed out. In a neutron star, matter is so tightly squeezed that just a teaspoonful would weigh 100 billion tonnes on Earth (about the same as a medium-sized mountain).

PLASMA GLOBES WERE INVENTED BY NIKOLA TESLA

PLASMA

When atoms are heated to extremely high temperatures, their electrons become so energetic that the nucleus can no longer hold on to them. This creates a soup of atomic nuclei and "free" electrons called a plasma. Because the electrons are "free", plasmas conduct electrical currents and

WHY IS **GLASS TRANSPARENT?**

Every atom of matter consists of a nucleus around which is a cloud of orbiting electrons. These electrons orbit the nucleus at different levels – like lanes on a motorway – depending on their energy. But electrons are not fixed in their orbits. If an electron travelling in the "slow lane" gains the right amount of energy, it will jump up an orbit (known as a quantum leap). If an electron loses energy, it will drop down an orbit. But not all atoms are equal – some need a bigger energy boost to get their electrons to change orbit.

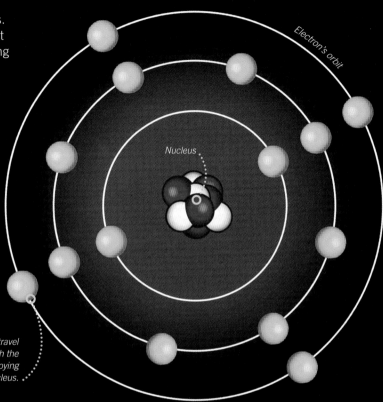

Electron's orbit

Nucleus

Light is made up of photons, which are tiny packets of energy – perfect for giving an electron a boost.

The more energetic electrons travel in the lanes further out, with the lower-energy electrons occupying the lanes closer to the nucleus.

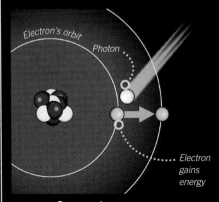

Electron's orbit

Photon

Electron gains energy

Opaque atom

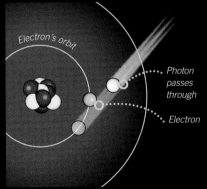

Electron's orbit

Photon passes through

Electron

Glass atom

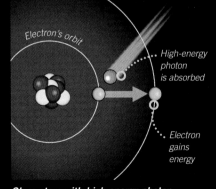

Electron's orbit

High-energy photon is absorbed

Electron gains energy

Glass atom with high-energy photon

NON-TRANSPARENT
When a photon comes into contact with the electrons of an opaque (non-transparent) substance, the electron gains enough energy from the photon to jump up a level in its orbit. The photon is absorbed – so light cannot pass through.

TRANSPARENT GLASS
In glass, it takes a lot more energy to make the electrons jump up to the next level (called the energy gap) than a photon can provide. The electron does not absorb the photon – so the photon of light can pass through.

ULTRAVIOLET
However, some photons are more energetic than others. Glass might be transparent to visible light, but higher-energy light (such as ultraviolet) has enough energy to kick glass's electrons up a level. This makes glass opaque to higher-energy light.

CURIOSITY:
SCIENCE'S HEART

HUMANITY IS AN INHERENTLY CURIOUS SPECIES. From the moment of our birth, we seek to understand ourselves, the world we inhabit, and all the space beyond. **Curiosity defines us.** The need to ask "what if?", "why?", and "how?" liberated us from the limits of an existence driven by survival alone and allowed us to become **the first species in the** history of the planet to try to understand how the world works.

Perhaps the ultimate expression of our curiosity is science. If curiosity is raw instinct, then science is curiosity channelled, focused, and refined – curiosity can survive without science, but science cannot survive without curiosity. **Remove curiosity from science and you tear out its beating heart.**

Yet scientific research can be costly, which has thrown up the argument that only scientific research that has a commercial value should be funded. Indeed, this argument believes that all other scientific endeavours are not valuable at all. In short, people who jump on this this train of thought want to **remove curiosity from science altogether.**

On the face of it, that might seem to make sense – after all, in these cash-strapped times, it is frivolous to fund money-pit projects like space telescopes when governments can back something that can be packaged, marketed, and sold for a profit. Nor is this a view limited to government policy-makers. Peer into the great World Wide Web and you will not have to dig too deep to find people asking such questions as **"Why spend billions on particle colliders or space telescopes when there are people dying of starvation, cancer, or war?"** Sure, if you look at great curiosity-driven science projects superficially, it may be difficult to see how they might benefit humanity beyond the "frivolous" quest for understanding – after all, how can £2 billion ($3 billion) spent trying to get a better view of a distant galaxy possibly have any effect on your daily life?

But the fact is, most of our modern world is built on the foundations of science driven only by "what if?", "why?", and "how?".

When Scottish physicist James Clerk Maxwell performed his experiments with electricity and magnetism in the late 19th century, **he was not aiming for something as base as personal profit or even anything as lofty as benefiting society.** Yet his electromagnetic tinkerings now form the foundations of our entire economy and society. Everything from computers, the Internet, satellites, mobile phones, and televisions to life-support machines, medical scanners, and machines that go "ping" owe their existence to science for curiosity's sake.

A century ago, when British scientist William Bragg investigated the strange patterns created by X-rays as they scattered from crystalline substances, he did not do it with the aim of creating a technique (X-ray crystallography) that would reveal the structure of DNA and revolutionize the fields of medicine, chemistry, physics, and engineering – **he did it out of curiosity and the desire to reveal something new about the way the world works.**

A more recent example is the discovery of graphene. Graphene is a material made of a single layer of carbon atoms that is 200 times stronger than steel and able to conduct electricity like nothing else. It is expected to replace silicon in microprocessors and should make it possible to build computers a hundred times faster than those today. Batteries made of graphene will charge hundreds of times faster than conventional batteries – **meaning a smartphone could charge in 30 seconds** and an electric car with a flat battery could be ready to drive away in minutes.

But graphene was not discovered by scientists looking to revolutionize electronics, but by two Russian guys at Manchester University playing around with sticky tape and a block of graphite – just to see how thin they could get it.

Examples like these are legion, but what would happen if Maxwell, for example, were to approach a research council today and **ask for funding "just" to see how something works?**

CURIOSITY AND ALBERT EINSTEIN

It was curiosity that drove Einstein to do his work. Yet, even without an economic goal, his work has made billions for companies and governments and helped build our modern world:

• His work on the photoelectric effect made it possible to build all the televisions, DVD and CD players, digital cameras, and remote controls ever sold.
• He created a formula to measure the size of molecules dissolved in liquids that has been used by chemists to create the shaving creams and toothpastes you use every morning.
• His theory of general relativity helps keep the atomic clocks used by GPS satellites synchronized so you do not drive off a cliff.

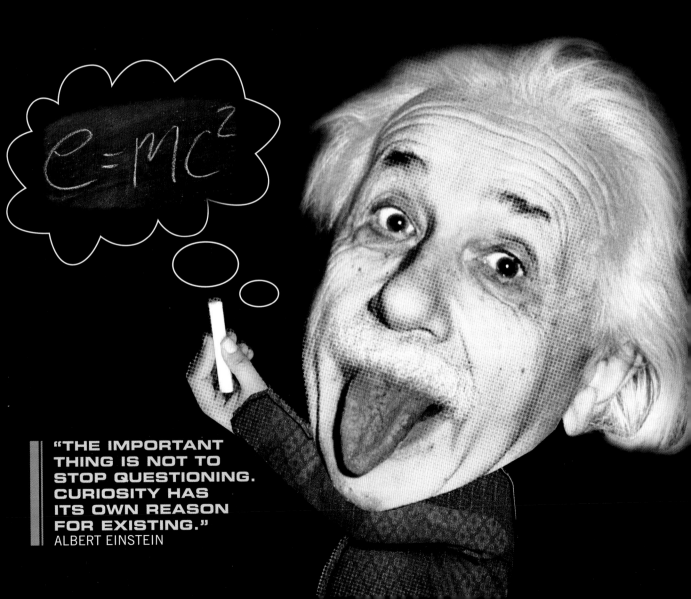

"THE IMPORTANT THING IS NOT TO STOP QUESTIONING. CURIOSITY HAS ITS OWN REASON FOR EXISTING."
ALBERT EINSTEIN

Well, if he were to ask those who seek only direct commercial returns from the world of science, he might very well be turned down. But does this law of unanticipated returns apply to all the sciences?

Much has been written about the spin-offs from large-scale physics research, so we thought we would explore those other cosmology mainstays – astronomy and space exploration. While it is quite easy to see how something like physics can have a long-term impact on our society, **it is perhaps more difficult to see how astronomy and space exploration could have much of an effect on those of us shackled to the surface of Earth.** Sure, it is hard to argue that discoveries of supermassive black holes and radiation-spewing neutron stars have much effect on Mr Joe Public, but in order to make those discoveries, **astronomers often have to invent new instruments and techniques that produce spin-off technologies that can (and do) have more tangible applications.**

And let us not overlook the power that **big science projects can have to inspire the next generation of scientists and engineers** to want to become that next generation. If we sideline curiosity and turn science into a just-for-profit enterprise, we will breed a generation of nine-to-five technicians and lose the Maxwells, Einsteins, and Braggs, who harness curiosity and create scientific revolutions.

SOME SPACEY
SPIN-OFF TECHNOLOGIES

Space exploration brings unique challenges that require solutions to be developed by engineers and scientists. Some find unexpected applications here on Earth.

> **"THE ALCHEMISTS IN THEIR SEARCH FOR GOLD DISCOVERED MANY OTHER THINGS OF GREATER VALUE."**
> ARTHUR SCHOPENHAUER (PHILOSOPHER)

WIRELESS TECHNOLOGY
Techniques developed in the 1970s to analyse signals from radio telescopes were adapted two decades later – by the same scientists – to reduce interference in radio-based computer networks. Wireless technology, commonly known as "WiFi", is now at the heart of modern wireless Internet communications.

Internet connections in Asia

SUNGLASSES
Without Earth's atmosphere to act as a filter, astronauts working in space are subjected to high levels of solar radiation – especially ultraviolet, which can cause permanent damage to the eyes. To protect astronauts' sight, NASA developed special coatings in the 1980s that blocked harmful radiation. Some commercial sunglasses use the same technology.

Coatings, designed to protect the sight of astronauts working in space, now look after yours when relaxing on a beach.

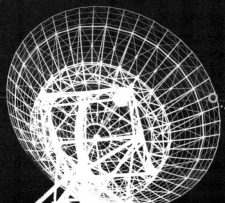

Wireless technology has helped us go from searching the heavens to searching the Internet.

In space, there are no plug sockets.

Driving on the Moon may not have been a game, but the technology has made computer games more fun.

CORDLESS TOOLS

Portable, self-contained power tools were originally developed to help Apollo astronauts drill for Moon samples. Back on Earth, this technology has led to the development of such tools as the cordless vacuum cleaner, power drill, shrub trimmers, and grass shears.

JOYSTICK CONTROLLERS

When NASA developed the Apollo Lunar Rover, they needed to develop an intuitive system that would allow the spacesuit-encumbered astronauts to steer and control the vehicle. They came up with the "joystick", which is used today to control everything from computer games to disability mobiles.

Cordless power drill

MEDICAL IMAGING

The Hubble Space Telescope needs to be super-sensitive to collect faint light from distant stars. The charge-coupled devices (CCDs), which convert light into digital files, were developed to meet Hubble's needs and have been adapted to greatly improve the sensitivity of digital biopsy machines used to detect breast cancer.

EAR THERMOMETER

The ear thermometer allows doctors to measure body temperature with accuracy and minimal invasion, but the technology started out as an astronomy tool. It uses a lens to detect infrared radiation (or heat) – in miniature, it is used to take your temperature but, in a telescope, it can detect the birth of stars.

Designed to take the temperature of stars, now it's taking yours!

Infrared thermometer

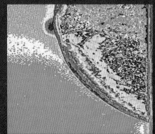

Microscopic image of biopsy done using Hubble-derived CCD technology

Hubble Space Telescope

Mathematical algorithms, used by astronomers to analyse all the data collected by telescopes, have been used to improve medical imaging.

FIREFIGHTING

Engineers have developed a water pump that allows firefighters to extinguish a blaze in one-sixth of the time it would have previously taken. The pump – which also uses just a fraction of the water – makes use of vortex technology designed to pump fuel into rocket engines.

WATER PURIFICATION

Water is essential to life, but it is much too heavy to be transported into space in large enough quantities for astronauts to live on. NASA developed filtration technologies that turned waste water from respiration, sweat, and urine into drinkable water. The technology is now being put to use in water-starved developing countries.

Water filter

THE CERTAINTY
OF UNCERTAINTY

THE WORLD OF
THE INSANELY
TINY

QUANTUM
GRAVITY

X-RAY CRYSTALLOGRAPHY

SEEKING
SUPERSYMMETRY

THE STORY OF THE ATOM

ATTACK OF THE MICRO  BLACK HOLES

TEENY, TINY, SUPER-SMALL
STUFF

HIGGS BOSON:
A BLUFFER'S GUIDE

DISCOVERING THE NEUTRON

PARTICLE ACCELERATORS

THE STORY OF **THE ATOM**

THE STORY OF THE ATOM BEGINS in ancient Greece in the 5th century BCE, when a tunic-clad thinker called Democritus formulated the idea that **matter might be comprised of tiny particles that were *atomos*, or indivisible.** These "atoms" could not be broken up because there were no particles any smaller.

For more than two thousand years, there was not much action in the story of atomic science, until finally, **in the early 19th century, an English physicist and chemist called John Dalton formulated his own "atomic theory".** Dalton's atom was much the same as the ancient Greek's, but he went on to suggest that the different elements were made of atoms of different sizes and that **the elements could be combined to create more complex compounds.** Dalton was also the first person to make a serious attempt to calculate the atomic mass of some of the chemical elements and to introduce a system of chemical symbols.

The next big step took place in **1897.** A British physicist, called Joseph John Thompson, was trying to figure out the nature of cathode rays – a mysterious blast of electromagnetic radiation emitted by a cathode (the negative part of an electrical conductor) within a vacuum tube. When he applied a positive charge, he noticed that the rays were attracted to it – **meaning that they must carry a negative charge.**

But the real breakthrough came when he calculated their mass and discovered that **they were about 1,800 times less massive than even the lightest atom** (hydrogen).

THE SEARCH FOR **STRUCTURE**

In the 19th century, scientists thought they had got to the heart of matter. But by the 20th century, ideas about atoms had been revolutionized and the challenge was to identify the structure of this puzzling particle.

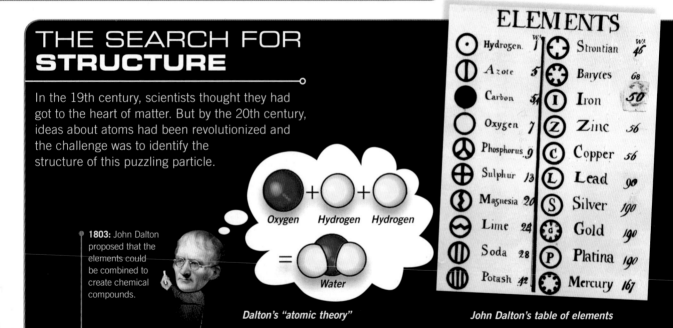

1803: John Dalton proposed that the elements could be combined to create chemical compounds.

Oxygen Hydrogen Hydrogen

= Water

Dalton's "atomic theory"

ELEMENTS

	w.			w.
⊙ Hydrogen.	1	✪ Strontian		46
⊘ Azote	5	✪ Barytes		68
● Carbon	54	Ⓘ Iron		50
○ Oxygen	7	Ⓩ Zinc		56
☮ Phosphorus	9	Ⓒ Copper		56
⊕ Sulphur	13	Ⓛ Lead		90
⊘ Magnesia	20	Ⓢ Silver		190
⊖ Lime	24	Ⓖ Gold		190
⊘ Soda	28	Ⓟ Platina		190
⦿ Potash	42	✪ Mercury		167

John Dalton's table of elements

1800 ❱❱ 1850

Since they were so small, he concluded that they must have come from inside atoms – **the indivisible atom must be divisible.**

Thompson called these tiny negatively charged particles "electrons" and incorporated them into a revolutionary new model of the atom. He knew that atoms are neutral (carrying no overall electrical charge) so, to balance out the negative electrons, **he imagined the atom as being a sort of cloud of positive charge peppered with electrons – like pieces of plum in a plum pudding.**

Although Thompson went on to win the Nobel Prize for Physics for his discovery of the electron, **his plum pudding model of the atom would only last about 10 years.** In 1909, a New Zealand-born physicist, Ernest Rutherford, was looking over the results of an

"IT WAS ALMOST AS INCREDIBLE AS IF YOU HAD FIRED A FIFTEEN-INCH SHELL AT A SHEET OF TISSUE PAPER AND IT CAME BACK AND HIT YOU"
ERNEST RUTHERFORD, DESCRIBING THE RESULTS OF THE GOLD FOIL EXPERIMENT

experiment performed by two of his students when he spotted a flaw in Thompson's atomic model. The students, Hans Geiger and Ernest Marsden, were experimenting with radiation by firing positively charged particles at a piece of gold foil. Based on Thompson's model of the atom, they had expected the particles to shoot virtually unimpeded through the positive cloud of the atom, which, although positively charged, should have been diffuse enough to allow the heavier particles to barge through.

Instead, they saw that, while many of the particles did pass

through, some were deflected and a very small number of the others bounced right back. This led Rutherford to conclude that **the atom must possess an extremely localized concentration of positive charge at its centre.**

Rutherford proposed that **the nucleus was made up of distinct units of matter that he called "protons",** and he placed Thompson's electrons into scattered orbits around the nucleus like planets orbiting the Sun.

Under Rutherford's new "planetary model", the atom was revealed to be made up almost entirely of empty

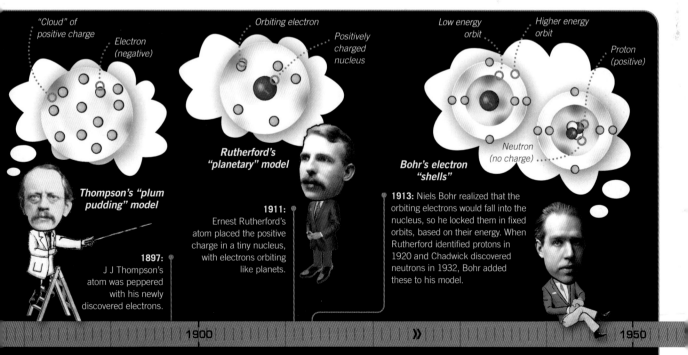

"Cloud" of positive charge

Electron (negative)

Thompson's "plum pudding" model

1897: J J Thompson's atom was peppered with his newly discovered electrons.

Orbiting electron

Positively charged nucleus

Rutherford's "planetary" model

1911: Ernest Rutherford's atom placed the positive charge in a tiny nucleus, with electrons orbiting like planets.

Low energy orbit

Higher energy orbit

Proton (positive)

Neutron (no charge)

Bohr's electron "shells"

1913: Niels Bohr realized that the orbiting electrons would fall into the nucleus, so he locked them in fixed orbits, based on their energy. When Rutherford identified protons in 1920 and Chadwick discovered neutrons in 1932, Bohr added these to his model.

1900   »   1950

MOST OF AN ATOM IS EMPTY SPACE – AN ATOM THE SIZE OF EARTH WOULD HAVE A NUCLEUS ABOUT THE SIZE OF A FOOTBALL STADIUM

space, with most of its mass concentrated in the tiny nucleus. But he had a problem: **what was stopping the negatively charged electrons from being pulled into the positively charged nucleus?**

To get around this, Rutherford dug into the kit bag of Newtonian physics and suggested that, just as the planets are kept in orbit by acceleration gained from the Sun's gravity, **electrons must be undergoing constant acceleration, which stops them falling out of orbit.**

Unfortunately, old-fashioned **Newtonian physics does not cut the mustard in the quantum world,** and this is where Niels Bohr enters the story. It was Bohr who saw the quantum flaw in Rutherford's otherwise ingenious atomic model.

A few decades earlier, Scottish physicist J C Maxwell had shown that when an electric charge is accelerated, it loses energy by emitting radiation (a process exploited by X-ray machines). Along with others, Bohr realized that Rutherford's accelerating electrons would lose energy by the same process and quickly fall into the nucleus. As this does not happen, **something else must be keeping an atom's electrons in check.**

On 6 March 1913, Bohr explained his modifications to the planetary model in a letter to Rutherford. Building on the work of German physicist Max Planck, who showed that there was a limit to how far something could move or be divided at the quantum level, Bohr

proposed that **electrons are restricted to fixed orbits depending on their energy.** Electrons with the least energy occupy the lowest orbit and those with the most energy occupy the highest orbits. **Electrons can only move between these orbits, or shells, by gaining or losing energy**.

But, in the mid-1920s, a whole new branch of physics, pioneered by the likes of Louis de Broglie, Erwin Schrödinger, and Werner Heisenberg, was entering the scene – quantum mechanics. In this weird new world, **the orbiting electrons became clouds of "possibility", in which they exist in all positions of their orbit at all times** (as both a particle and a wave) until observation forces them to assume position.

The last (slightly less weird) piece of the puzzle was revealed in 1932 when the British physicist **James Chadwick discovered a neutral partner to Rutherford's proton within the atomic nucleus: the neutron.**

Physicists at last had a pretty accurate model of the atom to work

NIELS BOHR

Danish physicist Niels Bohr won the Nobel Prize in Physics in 1922, and his model of atomic structure remains the basis of the physical and chemical properties of the elements. Like many of his generation, he was deeply affected by the events of World War II, and in later life he worked for the peaceful application of atomic physics.

with, but it was far from complete. **The indivisible atom that had been divided into protons, neutrons, and electrons would later be further divided into fundamental particles** (apart from the electron, which is as small as stuff gets). And so it was that the atom turned out to be more fascinating than anyone imagined.

ELECTRON CLOUD MODEL

By 1926, and the arrival of quantum mechanics, electrons were no longer thought of as orbiting particles but as waves. This led to a more abstract model of the atom, devised by the Austrian quantum physicist Erwin Schrödinger. He used the term "electron clouds" to describe where electrons were most likely to appear. This illustration shows a carbon atom with two electron clouds, coloured yellow and blue, to represent their different orbital paths.

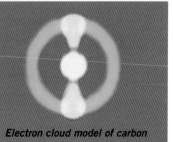

Electron cloud model of carbon

DISCOVERING
THE NEUTRON

THE NEUTRON IS HALF OF THE ULTIMATE DOUBLE ACT. It's the particle equivalent of Oliver Hardy – **the neutral straightman to the proton's charged personality**. Like all great double acts, the neutron and proton spent years plugging away in anonymity until, one day, they were discovered, plucked from obscurity, and thrust onto centre stage.

The proton and neutron can be **found at the heart of every atom** (apart from hydrogen, which possesses just one lonely proton), and **without them, matter as we know it could not exist.** Although the most fundamental of double acts, they did not find fame together. The proton enjoyed the first taste of celebrity, while the neutron was more reluctant to step into the limelight.

Rutherford knew there was something missing

Discovered in 1919 by New Zealand physicist Ernest Rutherford, the proton was initially encouraged to embark on a solo career as the only particle within the atomic nucleus. Like all great celebrities, the proton was known to be accompanied by a crowd of fans – known as the electrons. **The electrons were employed to keep the atom well balanced and neutral** (being negatively charged, they balanced out the proton's positive nature).

For a while, the arrangement seemed to work, but it soon became clear that **something did not add up.** The trouble was that an atom's atomic number did not always tally with its atomic mass – it was like there were more performers on stage than the billing had advertised.

To account for this discrepancy, Rutherford suggested that there might be **an as yet unseen performer at work within the atom,** another particle that had about the same mass as the proton

but, rather than being electrically charged, **possessed no charge at all.** This neutral particle would not upset the balance between the positive proton and the negative electrons. The hunt for the neutron was on, and the man to find it was Rutherford's assistant, British physicist James Chadwick.

But, having no charge, the neutron was rather difficult to locate. Fortunately, discoveries in Europe would provide just the trail of breadcrumbs that Chadwick needed to track the neutron down.

In 1930, researchers in Germany discovered that if you bombard the element beryllium with alpha particles (particles with two protons and two neutrons – like a helium atom but without the electrons), **a strange neutral radiation was emitted that could penetrate matter.** The discoverers of this phenomenon thought it was just common-or-garden gamma radiation, but Chadwick was not convinced and **believed that it was actually a particle**

WHAT'S MISSING?

On the periodic table, the atomic number refers to the number of protons found in the atom's nucleus. But the atomic mass is often more than twice that – meaning that the proton was not alone in the nucleus.

Atomic mass is double the atomic number

N	7
Nitrogen	

Atomic number

14

But his initial attempts to track down the particle in a cloud chamber (the usual method) proved fruitless. Then, in France, researchers discovered that if a lump of paraffin wax was placed in the path of the neutral radiation, **protons were knocked out.** To Chadwick, this was **proof a particle was at work.**

Anyone who has ever played (or watched) pool or snooker can understand why Chadwick came to this conclusion. Imagine the atoms within the paraffin are snooker balls. If you blow (our imaginary gamma radiation) on the snooker balls, you might succeed in moving a few of the balls but not much else. If you instead fire the cue ball (our neutrons) at the balls, you will see that **some balls are knocked out of the pack, just like the protons knocked from the paraffin atoms.**

Chadwick replicated the paraffin experiment, and he not only confirmed that **the neutral radiation was indeed a particle** but also, by tracing the paths and energies of the dislodged protons,

was able to figure out that **the particle must have about the same mass as the protons dislodged.**

At last, the neutron had been discovered and, as well as sharing the limelight with the proton, it went on to become a star in its own right. **The discovery of the neutron made possible the nuclear age.** Its ability to penetrate an atom's nucleus meant that it could be used to tear atoms apart and release the energy within (nuclear fission). Without Chadwick's discovery, there would have been no nuclear bomb (okay, so it's not all good) and no nuclear power stations.

Aside from helping to blow up Pacific islands, the neutron also has more benign talents, and **is an extremely useful tool for probing the atomic structure of matter.** Its ability to penetrate matter means it can tell us exactly where the atoms and molecules are within a material and how they behave.

If you think particle science is limited to the esoteric (such as what caused the Big Bang), you would be mistaken. At facilities like the Institut Laue-Langevin (ILL) in Grenoble, France, neutrons are used like supercharged X-rays to understand the world at the

JAMES CHADWICK

In June 1932, James Chadwick's paper announcing the discovery of the neutron was published by the Royal Society, and in 1935, he was awarded the Nobel Prize in Physics. As a result of his discovery, scientists across the world started bombarding all types of materials with neutrons. One such material was uranium and the result was nuclear fission. During World War II, Chadwick was the head of the British team working on the Manhattan Project in the USA, which led to the atom bomb.

WHAT'S INSIDE AN ATOM?

We used to think the atom was as small as things get – which is why the Greeks called them "atom", from *atomos*, meaning "indivisible". Atoms are made up of a nucleus (of protons and neutrons) and electrons, which orbit the nucleus. A neutron is (as its name suggests) electrically neutral, while protons carry a positive charge and electrons carry a negative charge. A neutron and proton are about the same size. They both dwarf the tiny electron (the mass of a proton is about 1,840 times that of an electron).

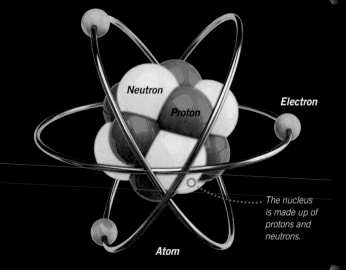

Neutron

Proton

Electron

The nucleus is made up of protons and neutrons.

Atom

atomic level. At ILL, the neutron has been used to develop **magnetic soap for mopping up oil spills, targeted cancer treatments, and new ways to combat viral and bacterial infections.** It has even helped make aircraft safer by finding structural defects hidden well beyond the reach of the human eye.

And how does ILL create the neutrons it uses? They have **their very own nuclear reactor** that feeds high-intensity beams of neutrons to an array of 40 instruments, which are used by some 1,200 researchers from over 40 countries every year.

The ILL is just one of the stages on which the neutron has performed in the 80 years that have allowed its meteoric rise from obscurity to be **one of the premier particle A-listers.**

> **CHADWICK CALCULATED THE MASS OF A NEUTRON AS 1.0067 TIMES THAT OF A PROTON, THUS SOLVING THE MYSTERY OF THE MISSING ATOMIC MASS**

HOW NEUTRONS **PENETRATE MATTER**

Having no charge, the anonymous neutron can pass undetected right to the heart of matter, where other particles fail.

1 ***Repellent protons***
Being positively charged, protons are repelled by the electrical forces in atomic nuclei. This means that protons are pretty rubbish at penetrating matter.

2 ***Slipping past***
A neutron's lack of electrical charge allows it to slip past an atom's charged field like an anonymous fan with a backstage pass. Neutrons can even pass through sheets of heavy metals, such as lead.

3 ***Strike!***
When a neutron collides with a nucleus, it can act like a cue ball striking a pack of balls and knock particles out of the nucleus.

4 ***No way out***
Sometimes the neutron will become trapped in the nucleus, transforming it into a heavier form of the same atom.

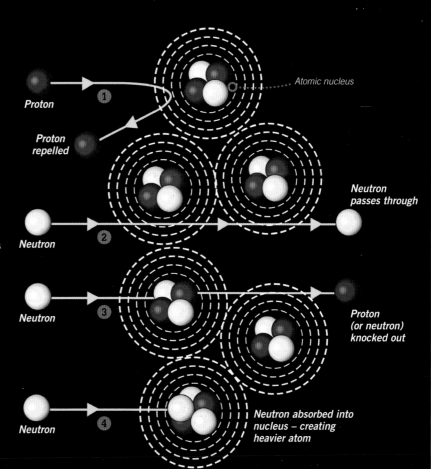

Proton

① Proton repelled

Atomic nucleus

② Neutron — Neutron passes through

③ Neutron — Proton (or neutron) knocked out

④ Neutron — Neutron absorbed into nucleus – creating heavier atom

THE WORLD OF THE
INSANELY TINY

THIS BOOK IS FULL OF INSANELY IMMENSE COSMIC STRUCTURES, such as stars, galaxies, and black holes – **stuff so large it literally bends the fabric of the Universe to its will.** We also like to indulge in a little brain blending by talking about **the world of the unimaginably small – the quantum world of the atomic and subatomic,** where our macroscopic view of reality is rendered impotent and the illogical reigns supreme. But beneath the quantum lurks another level of reality **where our ability to quantify reality breaks down and "small" takes on a whole new meaning.**

The Planck length is the limit to how small things can go

Too short to measure?

In theory, measuring extreme distances is restricted only by how far you are willing to go. **Keep doubling a distance and there is no limit to how far you can measure.** But does it work the other way round, if you measure the increasingly small?

It stands to reason that if you take a ruler and keep dividing it in half, there is no limit to how small the measure can go. But, as is so often the case in the quantum world, reason has precious little to do with it. Let us say you were to take a ruler (which for some bizarre reason measures 1.6 m (5.3 ft) in 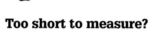 length) and divide it into 10 pieces. Now take one of those 10 pieces and divide that into 10 pieces, and so on. **In theory, you could repeat the exercise 34 times, but that is as far as you could go** and no force in the Universe enables you to divide it further. **This "last word" in measurement units is called the Planck length,** named after the father of quantum theory, physicist Max Planck.

The Planck length is 1.6 m (5.3 ft) divided by ten 35 times, a number with 34 zeros after the decimal point, which is really very small indeed and, as it turns out, as small as it is possible to go. **At this scale, the laws of physics we use to describe gravity, space, and time become useless.** If two somethings were to be separated by less than one Planck length, there would be no way to determine which something was where.

For the most part, we experience the Universe through interaction with the electromagnetic spectrum

(wavelengths of light, radio, and X-rays, for instance). Our eyes see where something is because they collect light photons that have interacted with the object (by bouncing off or being emitted).

All of the electromagnetic spectrum is transmitted by packets of energy called photons that have particular wavelengths – photons with more energy, like X-rays, carry

Just a tick

Time, theoretically, can tick onwards forever, but wind it back to its smallest increments – past the second, microsecond, and picosecond – and eventually you come to **the smallest possible increment: the Planck time.**

The Planck time is the smallest unit of time that can, in theory, be measured. One Planck time is the

According to quantum theory (in particular, Heisenberg's uncertainty principle), **if you cannot see it happen, then anything can happen** – a concept that naughty children and bankers sometimes try to apply to the non-quantum world. In this "grey" zone of accountability, particles of matter can "borrow" energy from the quantum vacuum and "pop" into existence literally from nowhere.

As long as the particles "pop" back out of existence and return their borrowed energy before the Planck time limit expires, the laws of "conservation of energy" (which

> ## "ANYONE WHO IS NOT SHOCKED BY QUANTUM THEORY HAS NOT UNDERSTOOD IT"
>
> NIELS BOHR, DANISH PHYSICIST

more energy and have shorter wavelengths than light photons. The shorter the wavelength of the photon, the smaller the object that photon can interact with – if it cannot interact, you cannot detect it. **The Planck length is so short that we could never create a photon with a short enough wavelength to interact with it** – therefore, we can never measure anything smaller. Even if we could create a photon with such a short wavelength, **it would carry so much energy, in such a small area, that it would collapse into a black hole** before it could return any useful information.

amount of time it takes a photon of light (travelling, naturally, at the speed of light) to cross a distance of one Planck length. One unit of Planck time is equal to about 10^{-43} seconds (or, 0.000000000000000000 00000000000000000000001 seconds) – **it is so short that there are more Planck times in one second than there have been seconds since the Universe began.** Anything that happens before the hands of the Planck clock move on by one unit is, by definition, unmeasurable, a quality that allows all sorts of quantum mechanical weirdness to take place.

Planck time splits the second to the point it becomes unmeasurable

MAX PLANCK

German theoretical physicist Max Planck was a professor at Berlin University, where he worked with his friend and collaborator, Albert Einstein. Between them, they were responsible for the two most revolutionary theories of 20th-century physics: Planck for quantum theory and Einstein for the theory of relativity. Planck was a talented musician. He sang and also played the piano, organ, and cello. Einstein sometimes accompanied him on the violin. In 1918, Planck was awarded the Nobel Prize in Physics.

state that energy cannot be created or destroyed) have not been violated. Perhaps the strangest outcome of the Planck time is that, because time cannot be measured within the Planck unit, **time as we think of it does not exist in the quantum realm.**

Since you and I are made of particles built of quantum "stuff", **time does not really "exist" as a tangible, measurable phenomenon for us either.** Even those who keep our clocks ticking as accurately as possible admit that **they do not measure time, they just define it.**

The seething vacuum

The discovery that space and time cannot be broken down beyond a certain point has implications for the way we understand the Universe. It shows that, because time and space each have a minimum dimension, at its most fundamental level **the Universe is built from tiny quantifiable units,** **or quanta,** which is where the science of "quantum" mechanics gets its name. Even the most featureless expanses of the Universe (the void, or the vacuum) are built from these quanta.

At the quantum level, "empty space" is never truly "empty", and the concept of a vacuum being a complete absence of something falls apart. **A vacuum just appears empty to us because there is no energy or matter that we can measure.** But beyond the measurable, in the quantum vacuum, **empty space is seething with virtual particles** that bubble up, live very (very, very, very) briefly on borrowed energy, and pop off again – something that physicists call "the quantum foam".

According to Planck, the quantum vacuum is full of virtual particles

HOW SMALL IS SMALL?

It would be impossible to show the Planck length to scale here, so let's start with an atom. This atom is 0.0000000001 m (0.0000000003 ft) in diameter (about 100,000 times smaller than anything you can see with the naked eye). If you were to try to measure the diameter of this atom in Planck lengths by counting one Planck length every second, it would take you about 10,000,000 times the current age of the Universe (that is 10 million times 13.8 billion years).

Planck length (10^{-35}) 1 yoctometre 1 zeptometre 1 attometre 1 femtometre 1 picometre 1 nanometre 1 micrometre 1 millimetre 1 metre

10^{-36} 10^{-33} 10^{-30} 10^{-27} 10^{-24} 10^{-21} 10^{-18} 10^{-15} 10^{-12} 10^{-9} 10^{-6} 10^{-3} 10^{0} m

The logarithmic scale

Atom

THE CERTAINTY OF UNCERTAINTY

QUANTUM MECHANICS IS ONE OF THE MOST SUCCESSFUL BRANCHES OF PHYSICS, when it comes to accurate explanations and testable predictions. It provides **the theoretical framework that allows scientists to describe how matter behaves at the subatomic level**. But despite its astonishing successes, quantum mechanics has an unfortunate side effect. It can induce the cerebral equivalent of **dropping a jellyfish into a blender** and transform the human brain into a quivering mess of gelatinous denial. To say that it is weird is an understatement of galactic proportions, and perhaps the weirdest of all quantum mechanics' predictions is something called **"Heisenberg's uncertainty principle"**.

Devised by genius German physicist Werner Heisenberg in 1927, the uncertainty principle states that, in the quantum world, **it is impossible to know simultaneously where a particle is and how fast and where it is going.** You can know its position or you can know its momentum, but you cannot know both. Okay, so perhaps that does not sound so very strange, but the reason behind it is very strange indeed.

Particles like electrons are not the discrete, spherical lumps of matter (like teeny, tiny ball bearings) we imagine them to be. In quantum mechanics, **a particle is a wavy smudge of spread-out potential.** It exists as a combination of all possible states, each state a combination of things like position, speed, and energy. Quantum mechanics perfectly describes particles using mathematical wizardry known as "wavefunction", which includes the likelihood of each state. However, when you want to make an observation of something, the system takes on one of the possibilities and the wavefunction collapses. It's a bit like rolling a die. As it scoots along the tabletop the numbers are a blur. Only by stopping the die can you "force" it to choose a number.

This indeterminate nature of the stuff that makes up the world around us did not sit well with scientists – after all, who wants to believe that the particles you are made of exist in state of quantum flux? **Even the physicists who created quantum mechanics were uncomfortable with its predictions,** which led Erwin Schrödinger to create his famous "cat in a box" thought experiment.

SCHRÖDINGER'S CAT

"Schrödinger's cat" is a thought experiment devised by Austrian physicist Erwin Schrödinger to demonstrate the absurdity of quantum mechanics. A cat is placed in a box with a vial of radioactive poison that will release a deadly gas when a single particle decays. Until the moment the box is opened and the state of the cat is revealed, the cat can be said to be both alive and dead at the same time.

Radioactive poison

Is the cat alive, dead, or both?

For decades, some scientists have expected (and hoped) **that uncertainty would one day be proved false** and that predictability would be returned to the Universe. But in 2013, their hopes were dashed by physicists working at the University of York, who published **a paper that would reinforce Heisenberg's description of the limits imposed by uncertainty.** By constructing a theoretical experiment, in which measurements of particles with known values were compared with those of particles whose states were unknown, they found that the errors in their measurements matched those in Heisenberg's original predictions.

Okay, so it was a lot more complicated than that, but **their conclusion could prove to be a boon for quantum cryptography.** Messages encoded in such a fashion would in theory be unbreakable because any attempt to "see" the message would force the multiple-state quantum bits that make it up to "collapse" to a single state (thus ruining the message).

> ## "I THINK I CAN SAFELY SAY THAT NO ONE UNDERSTANDS QUANTUM MECHANICS"
> RICHARD FEYNMAN, WINNER OF THE 1965 NOBEL PRIZE IN PHYSICS

SCHRÖDINGER

Austrian Erwin Schrödinger was one of the great theoretical physicists of the 20th century. Known to oppose Nazism, he fled his job at Berlin University, and was living in Oxford when he won the Nobel Prize in 1933 for his work on wave mechanics. He had a cat called Milton.

THE DECISIVE PARTICLE

Depending on how you look at it, light appears to have the properties of both a wave and a particle. When we test for wave-like properties, it behaves like a wave, but when we test for particle-like properties, it behaves like a particle. It seems to have multiple identities and exist in multiple locations at the same time. Can you feel your brain starting to quiver?

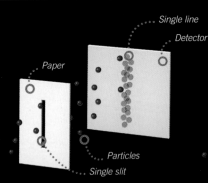

Paper · Single line · Detector · Particles · Single slit

1 A single slit
This is what happens when you shoot light particles (photons) through a slit in a piece of paper. As you would expect, they pass through the slit and leave a single vertical mark on a detector at the back – just as if you had fired a bunch of marbles through it.

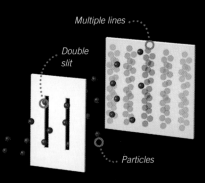

Multiple lines · Double slit · Particles

2 Double slit
So what happens if you make another slit? You would expect the particles to leave two matching lines, but, instead, there are many lines. How is this possible?

A CLOUD OF PROBABILITY

At the quantum level, matter does not really exist in a fixed state. Instead, it is a cloud of "probability" called the "wavefunction". This ability to exist in multiple states is called "superposition". Instead of thinking of a particle as a defined point of mass, it is more like a region of wavy potential smeared across space.

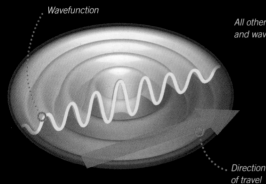

Wavefunction

All other probabilities vanish and wavefunction collapses.

Direction of travel

Particle's position

TRAVELLING WAVE
Like all waves, a particle wave is spread over a large area, so it has no definite position (peaks and troughs are regions where the probability of the particle occurring increase), but it does have direction. As a wave, we can know a particle's direction of travel, but we cannot know its position.

IN POSITION
In the same way, if we try to measure the position of the particle, the wavefunction collapses and we can no longer measure its direction of travel. This inability to measure all of a particle's properties is called "Heisenberg's uncertainty principle".

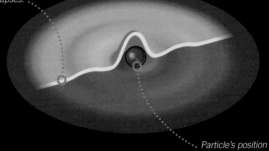

Two-line pattern shows particle-like behaviour

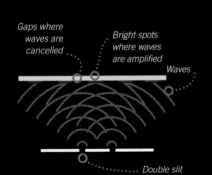

Gaps where waves are cancelled

Bright spots where waves are amplified

Waves

Double slit

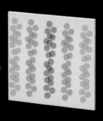

Beep!

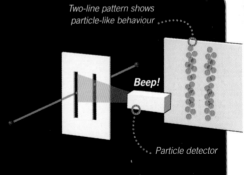

Particle detector

3 Wave behaviour
The only explanation is that the particles are behaving like a wave. When a wave passes through two slits, it spreads out from two fronts which then overlap. Where the peak of one wave meets the trough of another, they cancel each other out. Where two peaks meet, they amplify each other. This creates an interference pattern identical to the lines left by the particles.

4 Curiouser
Weirder still, if you fire just one particle at a time through the slits, you still get an interference pattern. This means that each single photon must have passed through both slits at the same time and then created an interference pattern as a wave.

5 And curiouser
If that was not weird enough, let us install a detector at the slits, which beeps every time a particle passes through one of the slits. It will see each particle pass through just one slit but the detector at the back will have just two lines on it. It is the same test as before, but this time there is no interference pattern. Baffling.

SEEKING
SUPERSYMMETRY

THE STANDARD MODEL OF PHYSICS, which describes the quantum world of particles and the stuff they are made of, **is one of the most successful theories in science**. Since it was first thought up in the 1960s and 1970s, it has made hundreds of predictions that have been successfully tested – the most recent of which was the discovery of the Higgs boson (the particle representative of the Higgs field, which imbues particles with mass).

Despite its success in the quantum realm, **the standard model (SM) only explains one aspect of the Universe** – gravity, space, and time do not fit. One theory that seeks to integrate the SM with the workings of the Universe at large is known as "supersymmetry" (SUSY). This is a collection of theories that predicts that, **for every particle in the SM, there exists a hidden, super-sized partner.**

LARGE HADRON COLLIDER	
Accelerator length:	27 km (16.8 miles)
Number of magnets: 9,300, cooled to -271°C (-456°F)	
Construction time: 14 years	
Initial cost: £3.6 billion ($6.4 billion)	

HIGGS HICCUP

Although the discovery of the Higgs boson supported the standard model, it also raised a new problem. Calculations based on the standard model predicted that the Higgs boson would have much more mass than it was discovered to have. A Higgs with a more massive "superpartner" (a Higgsino) would compensate for this. But the Higgs was also too "light" to fit with some supersymmetry predictions. One model predicts there should be five Higgs bosons (and related super-particles)... leaving physicists with one Higgs down and four to go.

Physicists are **hoping the Large Hadron Collider (LHC) will do for supersymmetry what it did for the Higgs boson.** After its highly successful initial run, the LHC underwent an upgrade in 2015 that saw its power double – and it will need every watt of that power to find a super-particle (sparticle). At their least massive, **these are predicted to be some 10,000 times more massive than a proton** – and they are likely to be much heavier.

Most sparticles are predicted to be highly unstable and, even if they are made in the LHC, **they will decay into a myriad of lighter particles within a fraction of a moment.** So, just as they did with the equally fleeting Higgs particle, physicists will have to pick through the decay debris in the hope that they can identify SUSY as the cause.

Like the hunt for the Higgs, the search for SUSY is likely to be a laborious process of **continually refining the energy band in which it may or may not be found** (the more massive the particle, the more energy is required to make it).

Even if the LHC fails to find anything after its upgrade, it will not mean SUSY is not out there, just that we are looking in the wrong place (energy) or for the wrong thing (it might decay into unexpected particles). But **if evidence remains elusive, physicists will face a tough choice** – to abandon the decades-old and heavily invested theory, or keep patching up a theory that may never yield direct evidence of its existence.

Whether SUSY's demise is mourned or not depends on who you talk to. For every physicist who sees an inherent beauty and elegant simplicity in the theory, there is another who sees her as **a Frankenstein-like patchwork of cobbled-together work around fixes.**

Collider upgrade:
The Large Hadron Collider is currently undergoing a comprehensive upgrade.

THE BUILDING BLOCKS OF MATTER (AND THEIR SUSY "COUSINS")

According to the standard model (SM) of particle physics, atoms are made of particles which, in turn, are made of fundamental particles. There are two families of fundamental particles – quarks and leptons. All matter is made up of a combination of two quarks ("up" and "down") and the lepton called the electron. The rest are usually only "seen" in high-energy particle collisions or in the moments after the Big Bang. Supersymmetry (SUSY) is an extension of the SM in which every fundamental particle has a "twin" (or partner) particle. If they exist, these super-particles (or sparticles) will have much more mass than their SM cousins.

FUNDAMENTAL PARTICLES

As far as we know, fundamental particles are the smallest building blocks of matter. They are divided into two groups: fermions (quarks and leptons) and bosons. Almost every particle has an antimatter version, identical except that it has the opposite electrical charge.

HOW BIG IS "MASSIVE"?

Based on the fact that the LHC has yet to turn up a sparticle at the energy it can search at, the lowest limit (for the least massive) would be about 10,000 times the mass of a proton (about the same as the difference between a mouse and a grand piano).

FORCE REACTIONS

Each of the fundamental forces of nature (electromagnetism, the strong nuclear force, the weak nuclear force) interacts with the fundamental particles through the exchange of force carrier particles called bosons.

THE STANDARD MODEL HAS MADE SO MANY SUCCESSFUL PREDICTIONS THAT IT IS OFTEN REFERRED TO AS "THE THEORY OF ALMOST EVERYTHING"

QUARKS

SQUARKS

Up

Down

Charm

Strange

Top

Bottom

Sup

Scharm

Stop

Sdown

Sstrange

Sbottom

LEPTONS

SLEPTONS

Electron

Electron neutrino

Muon

Muon neutrino

Tau

Tau neutrino

Selectron

Smuon

Stau

Selectron neutrino

Smuon neutrino

Stau neutrino

BOSONS

Higgs boson

W boson

Z boson

Photon

Gluon

Higgsino

Wino

Zino

Photino

Gluino

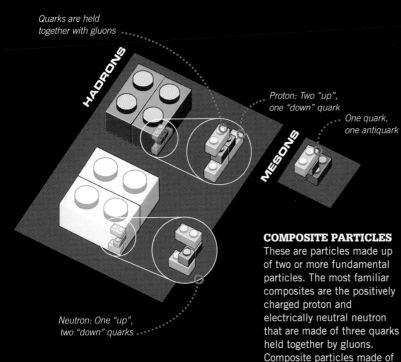

Quarks are held together with gluons

HADRONS

Proton: Two "up", one "down" quark

MESONS

One quark, one antiquark

Neutron: One "up", two "down" quarks

COMPOSITE PARTICLES

These are particles made up of two or more fundamental particles. The most familiar composites are the positively charged proton and electrically neutral neutron that are made of three quarks held together by gluons. Composite particles made of quarks are known as hadrons.

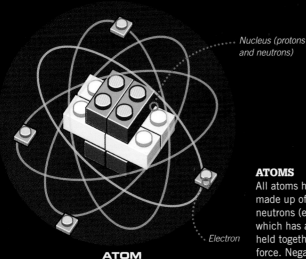

Nucleus (protons and neutrons)

Electron

ATOM

ATOMS

All atoms have a nucleus made up of protons and neutrons (except for hydrogen, which has a single proton), held together by the strong force. Negatively charged electrons orbit the nucleus. It is the electrons that allow atoms to bond together to create molecules.

QUARKS
• All of the matter in the Universe is made of a combination of "up" and "down" quarks.
• All particles composed of quarks are called hadrons (Greek for "heavy") – hence the name of the Large Hadron Collider.
• Quarks come in six "flavours", which have different properties and masses.

SQUARKS
The quark's more massive supersymmetry partner.

LEPTONS
• The most familiar lepton is the electron.
• Leptons are not made up of quarks (or indeed of anything smaller).
• The muon and tau are heavy electrons.

• Another lepton is the neutrino, a ghostly, almost massless particle that hardly interacts with matter.

SLEPTONS
The super-particle versions of the lepton – includes selectrons (left) and snuetrinos (right).

BOSONS
• Bosons are the particle messengers that tell other particles how to interact with the fundamental forces.

• The gluon mediates the "strong" nuclear force and is responsible for holding quarks together to form protons and neutrons.

• The photon is a tiny package of energy that carries the electromagnetic force, which affects any fundamental particle carrying a charge.

• W and Z bosons mediate the "weak" nuclear force, which is responsible for radioactive decay.

• Until recently, the Higgs boson was the missing piece of the SM. It is the particle representative of the Higgs Field, which gives mass to quarks and leptons (collectively known as fermions).

HIGGS BOSON:
A BLUFFER'S GUIDE

ON 4 JULY 2012, physicists at the European Organization for Nuclear Research (CERN), in Switzerland, home of the Large Hadron Collider (LHC), **announced the discovery of a new particle** that weighed in at about 125–126 GeV – that's about 130 times heavier than a proton. Two separate experiments had both detected the particle, with one data set achieving "five sigma" certainty (a one-in-3.5 million chance of error) that the particle was present. **The discovery was a vindication for the hugely expensive and massively ambitious LHC project,** built to find the mysterious particle. So what is the Higgs boson all about, and **how did scientists know what they were looking for before they found it?** Here is a "bluffer's guide".

LHC impact:
This conceptual artwork imagines the rays emitted from particle collisions in the LHC.

The Higgs boson was summoned into theoretical existence in the 1960s, to plug a gap in a theory that was almost perfect – the "standard model" of particle physics. The standard model has been highly successful. It can provide explanations and make predictions about how the counter-intuitive quantum world of physics works. But it couldn't explain one thing – why the fundamental particles have mass (it also can't explain dark energy and dark matter, but you can't have it all). According to the standard model, all the fundamental particles should have been born in the Big Bang without any mass at all. So how did the smallest building blocks of the cosmos summon mass as if from nowhere? The Higgs boson is seen as the answer to this problem. It is the physical emissary of an all-pervading field that interacts with fundamental particles to give them the mass we know they have.

THE **MISSING MASS**

According to the standard model, every object is made up of matter, which has mass. But the problem with this model is that it does not explain why the particle building blocks that make up matter do not have enough mass to account for the whole. It is like building a spaceship from six blocks, each of which has a mass of 1, and discovering that the total mass is 500. Something does not add up.

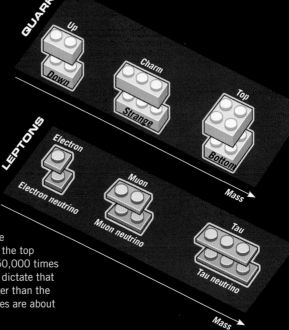

MASSIVE, MORE OR LESS
To add to the confusion, the standard model cannot explain why some particles have so much more mass than others. The lightest particle is the electron. The heaviest particle is the top quark, with a mass more than 350,000 times that of the electron. Logic would dictate that the top quark must be much larger than the electron, yet in reality the particles are about the same size.

Higgs believed particles acquired mass from a "force field"

WHY A BOSON?

A group of physicists, including Peter Higgs, proposed that the Universe is permeated by a sort of invisible force field. As particles travel through this "Higgs field" they interact with it and appear to acquire mass – the greater the interaction with the field, the greater their mass. We know from quantum theory that every field has an associated "force reaction" particle, called a boson, which acts like a messenger to transmit the effect of the field to the particle. So if there is a Higgs field, there must be a Higgs boson.

MASSIVE ATTACK – HOW HIGGS GIVES PARTICLES MASS

As with all things in quantum physics, the reality can only be described in abstract terms and complex mathematics. Not being quantum physicists, we will have to make do with an analogy – here using Elvis and his fans as the different particles, which of course is a huge oversimplification, but perhaps easier to grasp.

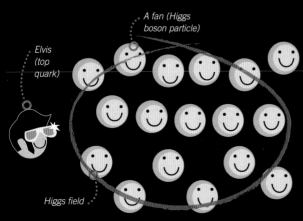

A fan (Higgs boson particle)

Elvis (top quark)

Higgs field

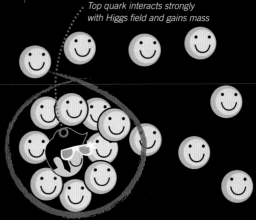

Top quark interacts strongly with Higgs field and gains mass

1 An even field

Imagine the Higgs field is a room filled with particle Elvis fans. The fans represent Higgs boson particles and are spread out evenly across the room.

2 Enter Elvis

Their hero, Elvis (representing a top quark), enters the room. The fans gather round and slow him down. He loses momentum and energy, but gains mass.

There is little interaction with the Higgs field

3 Much less attractive

When an Elvis impersonator (or electron) enters the room, the fans are not fooled and pretty much ignore him. With little to slow him down, he barely loses energy and gains almost no mass. In the same way, the more a particle interacts with the Higgs field, the more energy it loses, and the more mass it gains.

FIVE SIGMA

The Higgs boson discovery was given the "five sigma" seal of approval. Five sigma might sound like the title of a bad 1960s science-fiction movie but, for physicists, it represents success. A "sigma" is a measure of how likely it is that a result was down to chance. When physicists think they have found a reading that might be the Higgs boson, they repeat the experiment and look for the same reading. The more often they get the same result, the more the likelihood that it was a fluke is reduced. Only when the chance of it being a fluke is reduced to almost zero will they be able to say with confidence that they have found the Higgs. A five sigma result means there is only a one-in-3.5 million chance it is wrong.

SEARCHING FOR AN INVISIBLE
NEEDLE IN A HAYSTACK

To find evidence of the Higgs field, physicists look for the physical manifestation of that field – the Higgs boson. But they can't spot the Higgs boson directly, so instead they try to predict what sort of particles the Higgs will decay into and look for evidence of those. The action all takes place in the LHC.

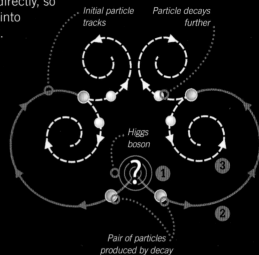

Initial particle tracks

Particle decays further

Higgs boson

Pair of particles produced by decay

1 Colliding protons

The LHC accelerates two beams of protons to 99.9999991 per cent of the speed of light (at this speed they complete 11,000 laps of the LHC's 27 km (16.7 mile) circumference every second). The collisions occur with so much energy that the physical laws that hold particles together (the "standard model") break down. Detectors trace and analyse the particles that emerge from the collisions.

2 Making tracks

Unfortunately, even if the collisions do spit out a Higgs boson, it will vanish almost as soon as it appears and decay into two different particles (in physics, decay means a particle turns into two lesser particles, not goes moldy and stinks out your fridge). Physicists study the tracks of these lesser particles.

3 New particles

To complicate matters further, when you smash two protons together, you get all sorts of particles being created – each of which will also decay. This image is a snapshot of one proton collision – all those lines and dots represent the particles that are created and their subsequent tracks. Imagine how difficult it is to find something in all that lot when you do not know exactly what you are looking for.

Red lines show the tracks of one pair of new particles

THE LHC IS
DESIGNED TO
GENERATE NEARLY
A BILLION PROTON
COLLISIONS EVERY
SECOND, WHICH
ARE ANALYSED BY
3,000 COMPUTERS

QUANTUM GRAVITY

ON THE FACE OF IT, TESTING HOW GRAVITY AFFECTS THE WORLD AROUND YOU seems like a straightforward proposition. All you have to do is pick something up – perhaps a cannonball, or a turtle – and then let it fall. Okay, not a turtle.

But **what if you are a physicist with a pocket full of subatomic particles** and you want to know how gravity affects these smaller-than-small

building blocks of nature? After all, **you cannot pick up a neutron with your quantum tweezers then drop it and expect to see what happens**. Well, you could travel to the Institut Laue-Langevin (ILL) in Grenoble, France.

At ILL, the physicists are neutron wizards who can literally bend particles to their will, and they are using their powers to **probe the mysteries of gravity within the quantum realm**.

NICELY CHILLED NEUTRONS

Compared to the strength of other forces, such as electromagnetism, gravity's power is so weak as to be almost undetectable. Most of the constituent parts of an atom, such as protons and electrons, interact enthusiastically with the other fundamental forces. So if you want to look at gravity, you need a particle for whom the other forces barely exist. The neutron is just such a particle. As its name suggests, it has no electromagnetic charge (neutral), making its interaction with gravity much easier to

spot. Like most particles, neutrons do not like to sit still. In fact, they zip around at thousands of kilometres per second, which is far too fast for any gravitational effects to be measurable in the lab. The best way to slow a particle down is to pop it into the freezer because the colder the particle is, the less energy it has and the slower it moves. At ILL, neutrons are chilled to within a whisper of absolute zero, which slows them to a much more manageable 9 m/second (30 ft/second).

Studying neutrons:
At the reactor at the Institut Laue-Langevin in Grenoble, neutrons are produced that can be used to probe the effects of gravity in the quantum realm.

Scientists at ILL have developed a groundbreaking technique, the first results of which were published in March 2014, to **slow down neutrons from their usual Concorde-like speed to a more sedate Usain Bolt-like pace.** At this speed, they can treat the neutrons a little like cannonballs and watch how they fall to Earth.

But **why do we need to know how something as small as a** **neutron interacts with Earth's gravity?** Quantum mechanics does a great job of telling us how stuff works in the world of the very, very small. Our theories of gravity (Newton's and Einstein's) do a top job of telling us how stuff works in the world of the large. But **we still do not know how gravity works at a subatomic scale.** The hope is that, by measuring how particles like neutrons interact with gravity,

physicists will be able to **unite quantum mechanics and gravity into a single theoretical framework.**

With rules for how particles usually interact with gravity, scientists can search for **anything unusual that might point to the existence of undiscovered particles** and forces, which might tell us more about one of the great mysteries in science today: **what are dark matter and dark energy made of?**

THE WEIRD WORLD
OF QUANTUM ENERGY

We are used to thinking of energy as being a sliding scale – you can have lots, little, or anything in between. There is no finite limit to how much you can have. In the quantum world of particles things are very different. Here, energy comes in discrete units, or quanta (hence "quantum"), which particles can absorb, or emit, to reach certain energy states – rather like rungs on an energy ladder.

Particle loses energy and drops to lower energy level.

Higher energy

Particle takes on energy and leaps up to higher energy level.

Low energy

POWER UNITS
You can think of energy quanta as being a little like batteries that particles use to power their climb up the energy ladder.

1 Moving up
A particle on the lowest rung can absorb one quantum battery and use its energy to leap to a higher rung – a "quantum leap".

2 Going down
A particle on a higher rung can shed a quantum battery to move to a lower rung.

STEP BY STEP
Just as you cannot cut a battery in half and expect it to work, a particle cannot absorb or shed less than one quantum unit. It cannot occupy a space between rungs on the energy ladder, any more than your foot can on a real ladder.

HOW TO MEASURE QUANTUM GRAVITY

You need to determine a scale before you can measure something – try using a ruler that has no centimetres or inches. For quantum systems, you need to be able mark your ruler with levels of energy states.

1 Cold start
At the ILL, a beam of cold neutrons travels above a polished glass plate. Each particle is full of gravitational potential energy that wants to fulfil its potential by falling to Earth.

2 The pull of gravity
As an object falls under the influence of Earth's gravity, its potential energy is converted into kinetic (movement) energy. When it bounces back up, kinetic energy turns into potential energy. When all the energy is converted, the neutron begins to fall back down.

3 Bouncing back
For the neutron, each bounce back up is like scaling the energy ladder. But with each bounce it loses energy, which means it cannot climb as high.

4 Minimum energy
If a neutron behaved like a ball, it would lose energy on each bounce until it simply rolled along. But because neutrons obey quantum laws, a neutron's "bounce" stops getting smaller when it reaches its minimum energy state.

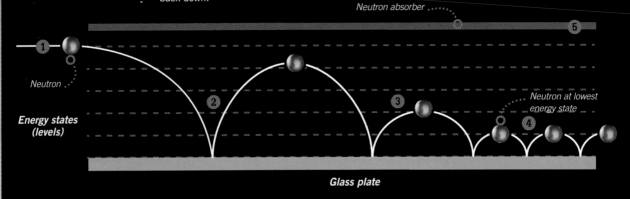

Neutron absorber

Neutron

Energy states (levels)

Neutron at lowest energy state

Glass plate

5 Neutron absorber
Above the neutron beam is an absorber that soaks up neutrons that strike it. As the absorber is lowered, it encounters neutrons at different energy states and absorbs them, reducing the number exiting at the far end. With each dip in neutrons, the energy level is noted. The point at which no neutrons make it out marks the lowest gravitational state.

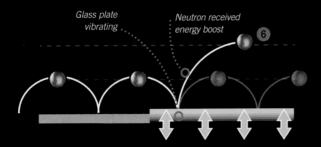

Glass plate vibrating

Neutron received energy boost

6 Energy boost
With only neutrons in their lowest energy state passing over it, the glass plate is made to vibrate. This adds energy to the neutrons so they jump to a higher energy state. Only if the plate vibrates at the right frequency will it add just the amount of energy needed to boost a neutron up. If too high or low, the neutron stays in its lowest energy state.

NEW PROJECTS

By measuring the frequency at which the neutrons jump up, physicists know what the energy difference is between the two states, allowing them to measure how much the neutrons are being affected by gravity and how much energy they are getting from the gravity field. This technique makes it possible to study with extraordinary precision how gravity operates in the quantum realm and determine if there are any as yet undiscovered forces at work. It will also allow physicists to search for evidence of new particles without having to rely on large, hugely expensive, "brute force" experiments such as the Large Hadron Collider.

X-RAY CRYSTALLOGRAPHY

IN 1913, BRITISH PHYSICIST WILLIAM HENRY BRAGG and his son, William Lawrence Bragg, made what is probably the most important discovery you've never heard of. They invented a technique, called X-ray crystallography, which allowed scientists to **look beyond the realm of the microscopic and into the kingdom of molecules and atoms,** revealing the hidden mechanisms that drive the world in which we live.

For the first time, **scientists were able to photograph atoms** by bombarding a crystallized sample with X-rays and then decoding the patterns left behind on photographic film.

Using X-ray crystallography, scientists could at last **decipher the hidden molecular structures that govern how materials behave** and figure out how the atoms within them interact. Almost overnight, the Braggs' discovery **revolutionized the fields of physics, chemistry, and biology.** And almost exactly 50 years later, X-ray crystallography was the key that **unlocked the mystery of the structure of DNA,** the code of life, ushering in the new science of molecular biology.

From biotechnology and pharmaceuticals to the planes we fly in and the fuels that power our planet, **there is virtually no area of our modern world that does not owe something to the discoveries of X-ray crystallography.**

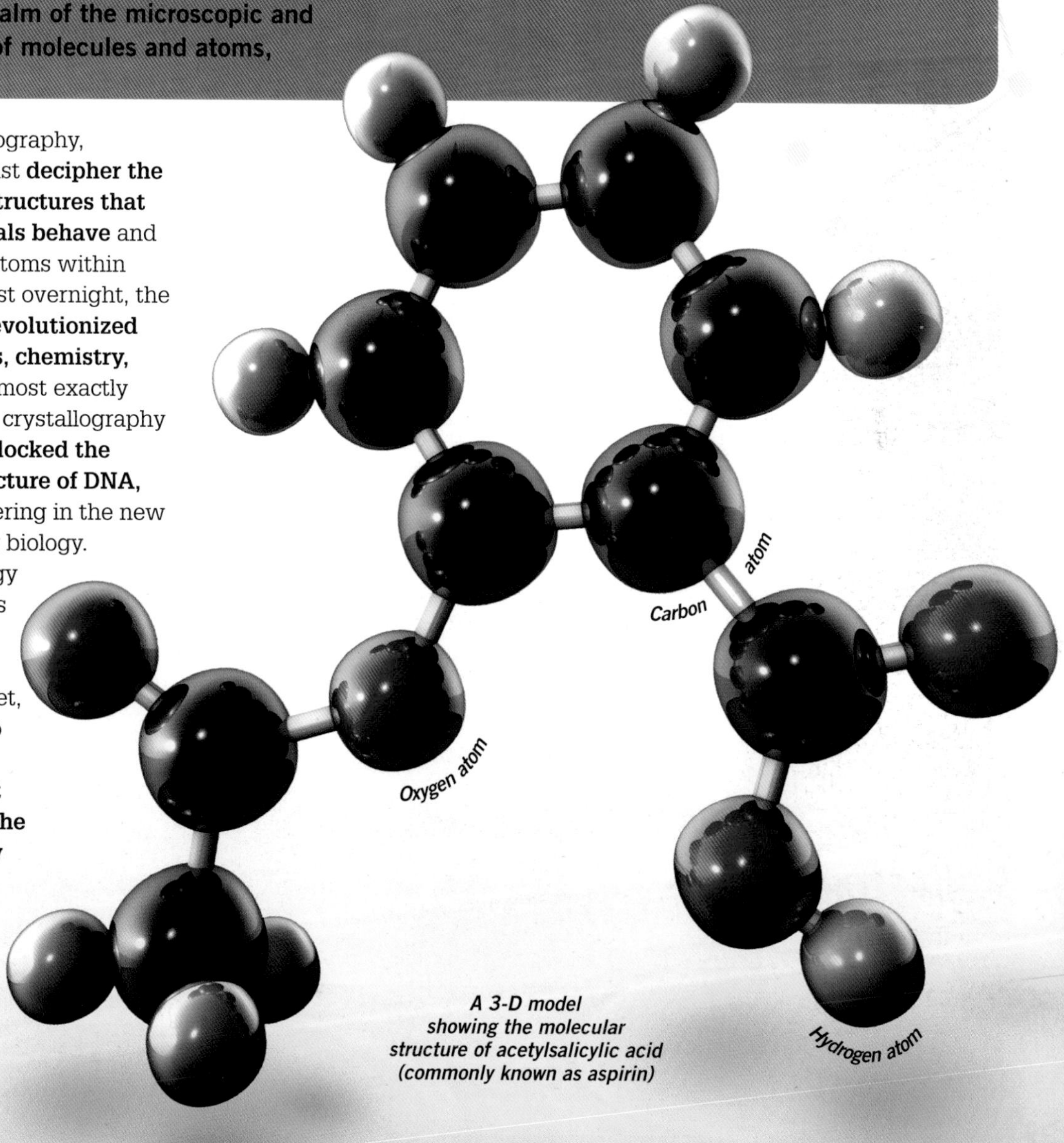

Carbon atom

Oxygen atom

Hydrogen atom

A 3-D model showing the molecular structure of acetylsalicylic acid (commonly known as aspirin)

HOW X-RAY CRYSTALLOGRAPHY
REVEALS HIDDEN STRUCTURES

Here's how X-ray crystallography makes it possible to "see" something as small as an atom. The sample to be studied must first be refined, purified, and concentrated to form a crystal. That's because in a crystal the molecules are organized into regular, repeating units, which make them easier to see. It's like the difference between trying to pick out a face from a milling crowd or an orderly queue.

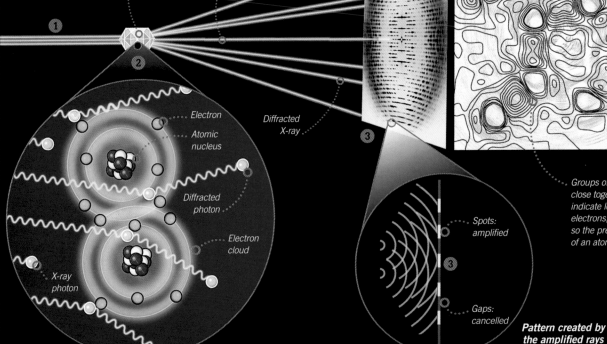

Diffraction pattern

Electron density map

Crystal sample

Undiffracted
X-ray

Electron

Atomic
nucleus

Diffracted
X-ray

Diffracted
photon

Electron
cloud

X-ray
photon

Spots:
amplified

Gaps:
cancelled

Groups of lines
close together
indicate lots of
electrons, and
so the presence
of an atom.

**Pattern created by
the amplified rays
on the detector**

1 · Light beam
A beam of X-rays is fired at the sample. Part of the electromagnetic spectrum (which includes visible light), X-rays are made up of packets (or particles) of energy called photons, but they also behave like waves.

2 · Diffraction
Most photons pass straight through the crystal, but the paths of some photons will be diffracted (made to change direction) as they strike the electrons in the atoms.

3 · Pattern of spots
The diffracted X-rays interact (or interfere) with each other. Some will be amplified and some cancelled out. The amplified rays will appear as spots on the detector, and these build up to create a pattern.

4 · Gradient map
As the spots were caused by photons diffracted by electrons, scientists can create a gradient map that plots how electrons are distributed within the sample. The higher the concentration of electrons, the closer the lines appear on the map.

THE BRAGGS ARE THE ONLY FATHER AND SON TO SHARE A NOBEL PRIZE. THEY ALSO HAVE A MINERAL NAMED AFTER THEM, CALLED BRAGGITE

Arrangement of atoms is plotted

3-D model of atomic structure

Atom

Chemical bond

AN AVERAGE GRAIN OF SAND HAS SOME 80 BILLION BILLION SILICA ATOMS – ALMOST CERTAINLY MORE THAN THE NUMBER OF GRAINS OF SAND ON THE BEACH IT CAME FROM

5 *Interpretation*
From the electron density, scientists can work out the position of atoms in the sample (where there are lots of electrons, there is an atom) and how they are bonded (through electron interactions). They can also work out which chemical element each atom belongs to (the higher the atomic number, the larger the electron cloud).

6 *3-D model*
By rotating the sample and taking images from different angles, scientists can build up a picture of the entire sample and construct a 3-D model of the molecule's complete atomic structure.

THE X FACTOR

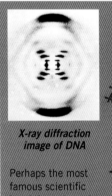

X-ray diffraction image of DNA

Perhaps the most famous scientific breakthrough made possible by X-ray crystallography was the discovery of the double helix structure of DNA. Above is the original "Photograph 51" that led the American scientist James Watson and the British scientist Francis Crick to make their Nobel Prize-winning discovery in 1953. The image, captured by Rosalind Franklin and Raymond Gosling in 1952, shows a distinctive "X" that Watson and Crick recognized as being the telltale sign of a helix.

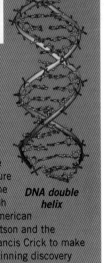

DNA double helix

PARTICLE
ACCELERATORS

THE LARGE HADRON COLLIDER
(LHC) at the European Organization for Nuclear
Research (CERN) is in many ways like a particle
sniper rifle. **It fires some of the smallest components
of matter into each other at colossal speeds with
exquisite precision,** so physicists can study the
even smaller components that come flying out.

There can be no doubt that **the LHC is a
machine in a class of its own.** It's the most
ambitious, technologically demanding, expensive,
and powerful particle sniper rifle ever built.
**But there is always a new model on the drawing
board** – a next-generation machine ready to
surpass its predecessor.

Despite its success in finding one
of its main targets, the Higgs
boson, in 2012, **the LHC was not
working at its full operating
potential at the time.** When
CERN's particle-colliding beast
was switched off for upgrades in
2013, it was only **running at a little
over half-power – eight trillion
electron volts (8 TeV).** When firing
on all 14 of its TeV cylinders, its
mission will be to find the answer
to **the big question for physics –
what is dark matter?**

Dark matter is so called because
it is invisible, or "dark", and its
existence can only be inferred
from its gravitational effect on

things we can see. **It is thought to
make up about 24 per cent of the
Universe,** so there is a lot of it.

The challenge is to find a
particle that has not been seen,
cannot be detected directly, could
exist in multiple forms (after all
normal matter is not made up of
just one sort of particle), and may
not exist at all. If that sounds like
an exercise in quixotic futility,
remember **they have done this
before with the Higgs boson.** But,
even operating at full power, the
mighty LHC may not be up to the
task. It might give some hints about
the nature of dark matter, and help
focus the hunt, but we may have
to wait for the next generation of
particle sniper rifles before
scientists are able to train their
sights on the elusive substance.

Proposals for a new machine
include building a giant linear
accelerator, or linac (perhaps the
most sniper-rifle-like accelerator),
which would hurl particles down
a **50 km (30 mile) long tunnel**
together with a sort of super-sized
LHC **with a 100 km (60 mile) long
accelerator ring.** Including
planning, it took **30 years and more**

than £5 billion ($8 billion) to build
the LHC. So, to build a more
ambitious machine than the most
ambitious machine ever built,
planning cannot start too early.

TYPES OF TRACK

LINEAR (LINAC)
A linac uses electromagnetic waves
to accelerate particles down a long,
straight track to collide with a target
(a magnetic field is used
to constrain the beam).

**CIRCULAR
(SYNCHROTRON)**
This accelerates
particles around a
circular track using
electromagnets.

*Linac feeds into
synchrotron*

COMBINATION
Most large particle
accelerators (like
the LHC) are a
combination of
linear and circular
accelerators.

DARK MATTER MACHINE

Here is a simple guide to the journey of protons through the LHC. Around the main ring are four areas – ATLAS, LHCb, ALICE, and CMS – where experiments are carried out on the speeding particles.

1 *Running start*
Protons set off at 33 per cent the speed of light in a linear accelerator.

2 *Speeding up*
In a booster ring they are bumped up to 91.6 per cent of the speed of light.

3 *Faster...*
The protons move into a 700 m (2,296 ft) synchrotron and are boosted to 99.93 per cent of the speed of light.

4 *Going underground*
They then shoot 40 m (131 ft) underground into a 7 km (4.3 mile) long ring where they are accelerated to 99.9998 per cent of the speed of light.

5 *The path splits*
Two streams of protons are fed into the LHC and circulate in opposite directions.

6 *Final surge*
To get that last 0.0001991 of a per cent closer to the speed of light, the protons are again boosted around the LHC's 27 km (16.7 mile) ring (covering about 11,000 laps of the ring every second).

7 *Impact!*
At 99.9999991 per cent of the speed of light, the two beams are smashed together within the four experiment areas.

8 *Making tracks*
In the energy maelstrom, all sorts of particle building blocks are created. Most are too short-lived to detect but, by tracing their characteristic tracks, physicists can infer the properties of the particles that created them.

ATLAS is a general purpose detector – it was used to hunt for the Higgs boson and will next explore micro black holes, dark matter, and extra dimensions of space.

Control room at ground level

Shaft to underground experiment area

ATLAS

LHCb

LHCb investigates matter and antimatter.

In ALICE lead atoms, rather than protons, are smashed together to recreate conditions after the Big Bang.

ALICE

CMS

CMS explores the same areas as ATLAS but uses different methods and technology, so results can be verified.

Tracks of post-impact particles

It is hoped that experiments will identify the dark matter particle.

MINI BIG BANGS

Physicists are not just looking for bits of broken proton kicked out by the impact – like shards of glass and metal thrown from a wrecked car. They are also looking for particles that have been created from the intense pressure and energy found in the dense subatomic fireballs (a million times hotter than the centre of the Sun) formed at the point of collision. This is why scientists talk about the LHC "recreating conditions at the time of the Big Bang": they really do create mini Big Bangs, in which the colliding protons melt into the same sort of hot and dense fundamental particle soup as the one from which the Universe emerged.

THE LOST **ELEMENTS**

Scientists will also be training the sights of their particle accelerators at other, more practical, problems, such as the chemical elements that make up the Universe.

The periodic table of elements is an icon – a list of atomic attributes that is as elegant as it is practical. As an at-a-glance guide to every chemical element, it is the scientific equivalent of the London Tube map. But it is far from complete. It is supposed to be the full list of all 98 naturally occurring chemical

elements (and 20 synthesized) but, according to some estimates, there are somewhere between 3,500 and 7,000 elements missing.

Now scientists are preparing to build two new particle snipers that will hunt down these "lost elements". To find them, scientists will have to recreate the violence of a supernova here on Earth. The first accelerator will be built in Germany, at the Facility for Antiproton and Ion Research (FAIR). The site of the second (whose acronym sounds like

a bladder medication), the European Isotope Separation On-Line facility (EURISOL), has not been decided. By smashing atoms of heavy elements, such as uranium, into each other (or into fixed targets), they will create temperatures more than a million times hotter than the Sun, and enough pressure, they hope, to produce some of the missing short-lived chemical elements, which they can then measure before they decay.

ILC: THE ULTIMATE **PARTICLE SNIPER**

The discovery of the Higgs boson in 2012 was a vindication for the expense of the LHC and a triumph of theoretical and experimental science, but it was not the end of the Higgs-hunting story.

Although the LHC did a fine job of finding the Higgs, the discovery raised more questions than it answered. For example, theoretical calculations predicted that the Higgs boson would have much more mass than the

particle discovered at CERN, raising the possibility that the LHC's Higgs is just one member of a larger Higgs family (of perhaps five Higgs).

Getting to know Higgs better will require the construction of a much more focused machine than the LHC. The International Linear Collider (ILC) will be much more precise than the LHC. Instead of

smashing protons together – which is a bit messy because they are made of smaller particles – the ILC will smash electrons into their antimatter equivalent, positrons.

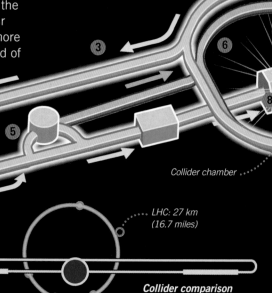

Collider chamber

ILC: 31 km
(19.2 miles)

LHC: 27 km
(16.7 miles)

Linear accelerator

Collider comparison

Uranium nucleus

New particles

MAN-MADE SUPERNOVA

FAIR will smash together uranium nuclei (the heavy radioactive element). The collision will create a fireball that briefly reproduces the extreme heat and pressure of a supernova explosion, creating around 1,000 new particles.

BURN-OUT

All the chemical elements heavier than iron are forged in the insane high-temperature, high-pressure conditions that exist when a star explodes as a supernova. Although the stable elements stuck around long enough to build the planets – and you and me – the vast majority were so unstable, they lasted just a trillionth of a trillionth of a second before they decayed into lighter, more stable, elements and were lost forever.

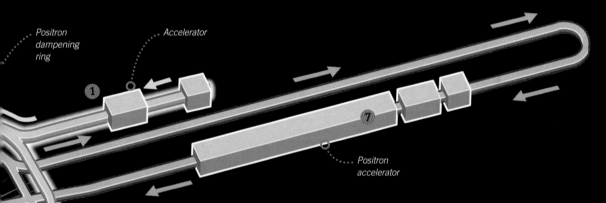

Positron dampening ring

Accelerator

Positron accelerator

Electron dampening ring

1 Electron entry
Bunches of 20 million electrons enter the accelerator.

2 Into the ring
The accelerated electrons are herded together into the dampening ring. In just a tenth of a second, the electrons complete 10,000 laps of the 6 km (3.7 mile) circuit.

3 Compression
When they leave, each bunch of electrons has been compressed into a beam just 0.3 mm (0.01 in) long and thinner than a human hair.

4 Acceleration
The beams are accelerated to 99.9999999998 per cent the speed of light in a linear accelerator (the LHC only manages a paltry 99.9999991 per cent).

5 Agitation
Some electrons are diverted and agitated by magnets until they emit photons. As the photons strike a titanium alloy target, the target's atoms emit more electrons and some positrons.

6 Rounding up
The positrons are corralled into focused bunches in their own dampening ring.

7 More speed
The positrons pass through their own 20 km (12.4 mile) long linear accelerator.

8 Smashing!
The electron and positron bunches are smashed together within a collider chamber. Detectors ten times more sensitive than the LHC's record the results.

ATTACK OF THE MICRO BLACK HOLES

BLACK HOLES ARE AMONG THE MOST EVOCATIVE and fascinating phenomena in the cosmos. Born from the collapsing cores of massive stars, they are the ultimate expression of gravity's power – **bending the fabric of space and time so absolutely that not even light can escape their clutches.** In their most massive incarnations, they lurk at the centre of every galaxy – **able to dictate the movements of stars and, if these stray too close, strip away their gaseous flesh.** Black holes are awesome and terrifying objects.

It is fortunate then that they can only be found in deep recesses of outer space... **but imagine if, during some sort of perverse science experiment, we were to make one here on planet Earth.**

When the Large Hadron Collider (LHC) was gearing up for its record-breaking high-energy particle collisions in 2008, **fear was rife that a deadly side effect of the collisions would be the creation of microscopic black holes.** Set free by the magnetic fields of the collider, the micro black holes would fall into the bowels of Earth where, nurtured in a womb of planetary material, they would grow, becoming increasingly massive **until they sucked up Earth and humanity along with it.**

Earth is still here, so it would appear the fear was unfounded. But it might surprise you to learn that some scientists hope that micro black holes will one day be detected in the aftermath of some of the LHC's particle collisions.

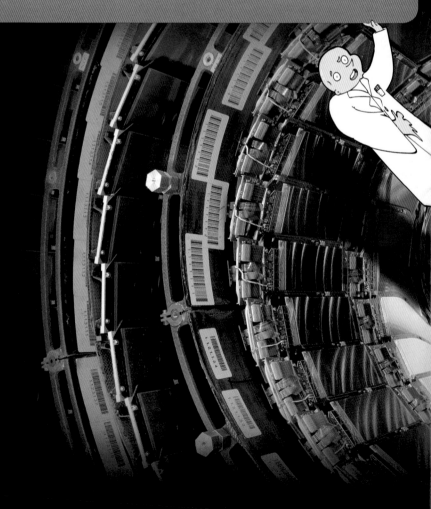

Lost in the LHC:
Some feared micro black holes would be created by particle collisions in the LHC.

You cannot be serious?

Why would anyone wish for something so scary? **It all comes down to the continuing search to understand gravity.** For big stuff, like the stars, planets, and you and me, gravity's effects are beautifully described by good, old-fashioned Newtonian physics and by Einstein's theory of general relativity. But for really small stuff, like atoms and their subatomic building blocks, gravity's effects stubbornly defy explanation.

The biggest problem is **gravity's apparent weakness compared with the other fundamental forces,** such as electromagnetism. Gravity might have the power to sculpt galaxies, but **the gravitational pull of an entire planet can be overcome by something as feeble as a child's magnet.** Under this framework, gravity is just too weak for even the most energetic of the LHC's particle collisions to result in the formation of a micro black hole.

But there is a theory that seeks to explain gravity's lack of muscle. **String theory predicts that, instead of there being just three dimensions of space, there might be as many as 26,** all curled in tightly bound knots and too small to be detected by our limited three-dimensional brains. The idea goes that, while the other fundamental forces are bound in three dimensions, gravity is free to roam all dimensions and, as such, becomes increasingly diluted.

When two particles collide at almost the speed of light, their energy is concentrated into a tiny space. **If extra dimensions do exist, gravity within that tiny space might be strong enough to allow the formation of a micro black hole.** If gravity can allow the formation of a micro black hole, then the next question is, **will a**

DEFYING GRAVITY

Gravity is so weak that you can overpower it yourself. Just grab a couple of nails and place them on a table. Gravity is using all its strength to pull the nails as close as possible to Earth's centre of mass. Then take a small magnet and watch in awe as its electromagnetic force easily dismisses the gravitational force of an entire planet and lifts the nails.

A magnet is all you need to defy Earth's gravity

Swallowed up?:
There's no danger Earth will
be swallowed up by a black
hole made in the LHC.

man-made black hole doom us
all? The short answer is no,
and here's why.

Micro means really tiny
Microscopic black holes are so
called because they are really,
really tiny. **Producing a black hole
is all about taking mass and
squeezing it until it falls below
the "Schwarzchild radius" – the
threshold at which gravity causes
the object to collapse in on itself.
You need a lot of mass to create
even a modest black hole – Earth
would squash down to a black
hole the size of a marble.**

In the LHC, the ingredients for
a potential black hole are in short
supply. **It would be created with
the mass of less than a couple of**

protons, so any resulting black hole
would be unimaginably small.

They won't devour the planet
The idea that **a matter-devouring
black hole would be tempted to
"fall" to the centre of Earth** is,
arguably, a logical conclusion, but it
is also wrong. If a teeny, tiny black
hole were to be created and then
liberated from the magnetic
confines of the LHC, it would be
travelling at much the same speed
as the particles from which it was
created. Since this is close to the
speed of light, **the black hole is far
more likely to shoot off into space.**
Whether the course of its planetary
exit takes it straight out into the
atmosphere, or through Earth's
core, is largely irrelevant. It's so

inconceivably small, **it would take
longer than the current age of the
Universe to devour just a gram of
our precious planet.**

They evaporate really quickly
Time and velocity are the least
of the obstacles faced by micro
black holes with planet-devouring
ambitions. **The greatest hurdle is
their insanely short lifespans.**

Black holes are famed for their
"everything in, but nothing out"
nature, so you would be forgiven
for thinking that a black hole of any
size could only ever get bigger. But
British physicist Stephen Hawking
says otherwise. In 1974, he realized
that all black holes actually emit
radiation, now known as Hawking
radiation, which causes them to

VIRTUAL PARTICLES

So-called "virtual particles" are born with an energy debt that they have to repay before the Planck time limit expires. Normally they do this by annihilating each other in a flash of energy that repays their quantum vacuum debt.

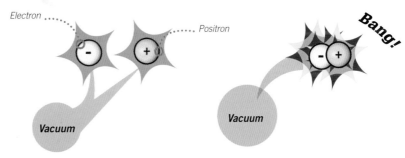

Electron
Positron
Bang!
Vacuum
Vacuum

1 **Pop-up particles**
A virtual particle pair are created, owing energy to a vacuum.

2 **Over in a flash**
Usually, the particles annihilate each other and release energy back to the vacuum.

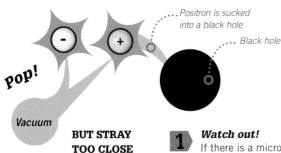

Pop!
Vacuum
Positron is sucked into a black hole
Black hole

BUT STRAY TOO CLOSE TO A MICRO BLACK HOLE...

Virtual electron becomes real
Black hole
Vacuum

1 **Watch out!**
If there is a micro black hole nearby, it sucks up a virtual particle, which is then lost to space and time.

2 **Energy debt**
The black hole "owes" energy to the vacuum and loses energy. As the virtual electron becomes real, it makes the black hole radiate energy.

lose energy continuously, which leads to their evaporation.

But, **if matter and energy cannot escape a black hole, how does it lose energy and matter?** Well, it comes down to a peculiar quirk of quantum mechanics. This tells us that **empty space is never truly empty.** On the smallest scales, it is a bubbling sea of quantum fluctuations from which pairs of particles can be created seemingly from nothing.

Heisenberg's uncertainty principle tells us that the shorter the amount of time you look at something, the less certain you can be of what is going on – **"if you cannot see it happening, anything is possible".** In quantum physics, the shortest period of measurable time is called "Planck time". Anything that happens within that time is, by definition, unmeasurable. If this is so, the uncertainty principle tells us that nothing is impossible.

With no "rules" to prevent it, particle pairs (electrons and their antimatter opposites, positrons) can "borrow" energy from the vacuum and pop into existence (see panel, above). But, if they are created too close to a micro black hole, one of the pair can be "sucked up" and lost to space and time. At that point, the remaining particle is forced to become a bona fide particle, and the black hole is left saddled with the energy debt of the particle it swallowed. Since debt is always a negative value, it effectively takes on negative energy, which is subtracted from the energy it has stored away. In this way, the micro black hole radiates energy (carried away by the virtual particle made real) and it evaporates too quickly to swallow even a small scientist in the LHC, let alone planet Earth.

THE EXPECTED LIFESPAN OF A MICRO BLACK HOLE IS LESS THAN ONE OCTILLIONTH OF A NANOSECOND

INDEX

ACKNOWLEDGMENTS

DK WOULD LIKE TO THANK:
Victoria Pyke for proofreading,
Carron Brown for the index, and
Bharti Bedi for editorial assistance.

The publisher would like to thank the following for
their kind permission to reproduce their photographs:

(Key: a-above; b-below/bottom; c-centre; f-far;
l-left; r-right; t-top)

2-3 Robert Gendler: (b). 6-7 ESA / Hubble:
NASA / http://creativecommons.org/licenses/
by/3.0/. 6 ESA / Hubble: NASA (br) / http://
creativecommons.org/licenses/by/3.0/. 7
Dreamstime.com: Danang Setiawan (c). NASA:
ESA / S. Beckwith(STScI) and The HUDF Team
(tr). 8 Adam Block/Mount Lemmon SkyCenter/
University of Arizona (Board of Regents): (cl).
NASA: (tr); GSFC; ESA / ASU / J. Hester (c);
HST (cr); ESA / STScI / A. Nota (bl); JPL (bc);
CXC / IoA / S.Allen et al (br). 9 Andrew Z. Colvin:
(bl, br). Lowell Observatory Archives: Jeffrey
Hall (tl). NASA: (ftl, tr, cr); STEREO (tc). Two
Micron All Sky Survey, which is a joint project
of the University of Massachusetts and the
Infrared Processing and Analysis Center/
California Institute of Technology, funded
by the National Aeronautics and Space
Administration and the National Science
Foundation: (bc). 10-11 Alamy Images:
ClassicStock. NASA: GALEX, JPL-Caltech (bg).
12 Getty Images: New York Times Co. (clb).
NASA: ESA and the Hubble Heritage Team
(STScI / AURA) (br). 14-15 NASA: GSFC.
17 The Library of Congress, Washington DC:
(br). 19 ESA: Planck Collaboration (br). 20 NASA:
Beckwith (STScI), Hubble Heritage Team, (STScI /
AURA), ESA (bl). 25 NASA: (tr). 26 NASA: JPL-
Caltech / UCLA (crb). 27 ESO: (b) / http://
creativecommons.org/licenses/by/3.0/. NASA:
JPL (tr). 28 NASA: JPL (bl); STEREO (cb/used
3 times in the spread). 29 iStockphoto.com:
Andy_R (bc). NASA: CXC / M. Weiss (tr/used 4
times in the spread). 30 Courtesy Jim Misti (bg).
NASA: JPL-Caltech (b). 31 ESA: NASA / SOHO
(clb). NASA: JPL-Caltech / Univ. of Ariz. (tr); JPL-
Caltech (bc, br, bl). NASA: (cla); R. Hurt (cra,
fcra). 33 ESO: (bg) / http://creativecommons.org/
licenses/by/3.0/. NASA: (cb, bl). 34 NASA: Johns
Hopkins University Applied Physics Laboratory /
Carnegie Institution of Washington (bc). 34-35
Nicolle R. Fuller, National Science Foundation:
(c). 35 NASA: (tc). 36 Dreamstime.com: Anton
Brand (fcl); Lineartestpilot (cl). NASA: JPL-
Caltech / UMD (b). 37 Dreamstime.com: Anton
Brand (fcra); Realrocking (cra). 38 ESA: MPS /
UPD / LAM / IAA / RSSD / INTA / UPM / DASP /
IDA (cra). NASA: JPL-Caltech / Cornell (ca); JPL-
Caltech (clb). 38-39 ESO: E. Slawik / http://
creativecommons.org/licenses/by/3.0/. 39 ESA:
Halley Multicolor Camera Team, Giotto Project
(clb). NASA: STEREO (cra). 40-41 NASA: JPL /
Space Science Institute. 41 NASA: JPL /
University of Colorado (cl); RSS, JPL, ESA (tr);
JPL / ESA (cr). 42 Dreamstime.com: Ncomics (tr).
NASA: (bl). 43 Dreamstime.com: Dedmazay (cra).
ESO: M. Kornmesser / Nick Risinger (tr) / http://
creativecommons.org/licenses/by/3.0/. 45 NASA:
(cla, ca); Ames / JPL-Caltech (br). 47 ESO: (bg) /
http://creativecommons.org/licenses/by/3.0/.

NASA: ESA / M. Kornmesser. 48 ESA: NASA /
SOHO (cra); University of Bern (bl). NASA: ESA /
Digitized Sky Survey 2 (tr); ESA / M. Kornmesser
(cl). 49 ESA: NASA / SOHO (c). NASA: ESA / M.
Kornmesser (clb, cb, cr). 50-51 NASA: CalTech.
51 NASA: JPL (br). 52 NASA: ESA, and A. Feild
(STScI) (b). 53 Dreamstime.com: Emily2k (tr). 56
Rex Features: Sovfoto / Universal Images Group
(b). 57 Science Photo Library: Ria Novosti (t). 58
Getty Images: Sovfoto (cl). 59 NASA: JPL (bc).
60-61 NASA: (r). 61 NASA: (cl). 62 NASA: (cb, ftr,
br, cb/pluto); The Hubble Heritage Team (STScI /
AURA) (tr); Goddard Space Flight Center (cra/
earth). 63 NASA: (ca, cra, bl, cla/Mars); JPL (cla).
64-65 ESO: (bg) http://creativecommons.org/
licenses/by/3.0/. NASA: JPL (clb). 65 NASA: (cr);
JPL-Caltech (fcr). 66 NASA: (cr, cla); Voyager 1
(cb). 67 NASA: (cla/Mars, clb/pluto); The Hubble
Heritage Team (STScI / AURA) (cl); JPL (ca, cb,
bl, br, cla/Venus, tr). 68-69 NASA: JPL / MSSS (t).
68 Dreamstime.com: Tranz2d (tr). NASA: (bc);
Goddard Space Flight Center (cla/earth). 69
Corbis: Science Picture Co. (crb). Lowell
Observatory Archives: (cra). NASA: D. McKay
(NASA / JSC), K. Thomas-Keprta (Lockheed-
Martin), R. Zare (Stanford) (bl). 70 NASA: JPL /
UA / Lockheed Martin (t). 71 NASA: JPL-Caltech
(t). 72-73 NASA: JPL. 73 NASA: (b, tr); Viking
Project, JPL (ca). 74 Dreamstime.com:
Maya0851601054 (cra); Sebastian Kaulitzki (tl).
NASA: (bl, bc, crb). 75 Dreamstime.com:
Sebastian Kaulitzki (tc). NASA: (cra, bl, cb).
76 ESA: D. Ducros, 2013. 76-77 NASA: NASA /
HST / CXC / ASU / J. Hester et al (Background).
78-79 ESA: P.Carril. ESO: (bg) / http://
creativecommons.org/licenses/by/3.0/. 79
NASA: Goddard / University of Arizona (br). 80
Dreamstime.com: Emmanuel Carabott (clb);
Gennady Poddubny (c); Jroblesart (ca); Peter
Hermes Furian (br). 80-81 ESA: P.Carril. ESO: (bg)
/ http://creativecommons.org/licenses/by/3.0/.
81 Dreamstime.com: Alexandr Mitiuc (tl);
Emmanuel Carabott (cr, fbr); Yudesign (tc). ESA:
P.Carril (c/asteroid). 82 NASA: JPL-Caltech /
Space Science Institute (clb). 83 NASA: Cassini
Imaging Team, SSI, JPL, ESA (br); JPL / Space
Science Institute (c). 84 NASA: Goddard Space
Flight Center (bl); JPL / Space Science Institute
(cr). 85 NASA: Cassini Imaging Team, SSI, JPL,
ESA (br); JPL / Space Science Institute
(Reproduced Fives Times On The page). 86-87
ESA: Northrop Grumman. 87 NASA: GSFC (ca).
88 ESA: ATG medialab (t). 88-89 ESO / http://
creativecommons.org/licenses/by/3.0/. 89 ESA:
ATG medialab (tl); Rosetta / MPS (br). 90 ESA:
(fclb, bl, clb/Steins); MPS / UPD / LAM / IAA /
RSSD / INTA / UPM / DASP / IDA (crb). 91 ESA:
(cl); ATG Medialab (cb/used 4 times); OSIRIS
Team MPS / UPD / LAM / IAA / RSSD / INTA /
UPM / DASP / IDA (bl). 93 Alamy Images: The
Stocktrek Corp / Brand X Pictures (ca). NASA:
(bc); JPL (c); ESA, STScI (br). 94 Dreamstime.
com: Eti Swinford (bl). 96 NASA: (bl). 97 NASA:
(cl). 100 Corbis: Heritage Images (bc); The Print
Collector (tr). Dreamstime.com: Suljo (bc/paper).
101 Corbis: Louie Psihoyos (br). 102-103 Science
Photo Library: Pasieka (bc). 103 Corbis:
Bettmann (tr). 108 Alamy Images: Thomas
Henrikson (c). ESO (bg) / http://creativecommons.
org/licenses/by/3.0/. 109 National Physical

Laboratory: (crb). 110 ESO: M. Kornmesser (cl) /
http://creativecommons.org/licenses/by/3.0/.
111 ESA: Planck Collaboration. 113 ESA: Planck
Collaboration (cr). 114-115 Millenium
Simulation: Springel et al. Nature 435, 629
(2005).. 115 Science Photo Library: Emilio Segre
Visual Archives / American Institute of Physics
(cra). 117 NASA: (tc). 118 NASA: (b). 120 S.
Cantalupo 2014: (t). 121 S. Cantalupo 2014: (br,
cr). 123 NASA: ESA, and the Hubble Heritage
Team (STScI / AURA) (b). 127 iStockphoto.com:
Andy_R (cr). 128-129 ESO: L. Calçada (b) / http://
creativecommons.org/licenses/by/3.0/. 129 ESO:
L. Calçada (ca, cra, fbr) / http://creativecommons.
org/licenses/by/3.0/. NASA: JPL (br). 130-131
ESO: L. Calçada (c) / http://creativecommons.org/
licenses/by/3.0/. 130 Dreamstime.com: Aleksey
Mykhaylichenko (fbl). ESO: L. Calçada (bl, cra) /
http://creativecommons.org/licenses/by/3.0/. 132
NASA: JPL-Caltech. 136-137 NASA: Goddard
Space Flight Center (b). 138 NASA: ESA, and M.
Livio and the Hubble 20th Anniversary Team
(STScI) (tr). 139 NASA: (br); GSFC (cb). 141
NASA: ESA / ASU / J. Hester (cr); JPL (tl);
SDO (bc). 142 NASA: Image processing by R.
Nunes http://www.astrosurf.com/nunes (c).
142-143 ESO. Richard Kruse: (b) / http://
creativecommons.org/licenses/by/3.0/. 143
Wikipedia: (cr). 144 ESA: (b); NASA, A. Simon
(Goddard Space Flight Center) (crb/used 8 times
in the spread). NASA: JPL-Caltech (tr/used 4
times in the spread); STEREO (cb/used 4 times
in the spread). 146 Alamy Images: fStop Images
GmbH. 148 Corbis: Danilo Calilung (bl).
Dreamstime.com: (cra/Flask); Lineartestpilot
(cra); Rafael Torres Castaño (c); Oguzaral (cb);
Martin Malchev (crb). 151 Alamy Images:
Interfoto (br); Simon Belcher (cl). 152 Alamy
Images: Photopat (bl). Getty Images: Fry Design
Ltd (cb). 152-153 Dreamstime.com: Svsunny.
153 NASA: (clb). 156 Corbis: Bettmann (crb).
The Library of Congress, Washington DC: (bl).
157 Corbis: (cb); Bettmann (br). Getty Images:
Print Collector (bl). 158 Science Photo Library:
David Parker (br). The Library of Congress,
Washington DC: (tr). 159 Corbis: Bettmann (cr).
Fotolia: valdis torms (crb). 160 Fotolia: valdis
torms (br). Getty Images: Elliott & Fry / Stringer
(c). 162 Alamy Images: Photo Researchers (cla).
Dreamstime.com: Alexander Kovalenko (c). 163
Alamy Images: Photo Researchers (cr). Corbis:
Bettmann (bl). Dreamstime.com: Lineartestpilot
(crb). 164 Alamy Images: Photo Researchers (tr).
Dreamstime.com: Andrei Krauchuk (cr); Elena
Torre (crb, bl). 165 Dreamstime.com: Liusa (br);
Xcenron (bc); Shtirlitc (bc/flask, fbr). 166 Corbis:
Bettmann (c). Dreamstime.com: Xcenron (cl).
168-169 © CERN : Maximilien Brice. 172 Getty
Images: PASIEKA. 173 Corbis: Martial Trezzini /
epa (cl). 175 © CERN : ATLAS, Collaboration (br).
176-177 ILL: JL Baudet. 179 Science Photo
Library: Animate4.com. 181 Jerome Walker: (fcr).
Science Photo Library: (cra). 186-187 © CERN :
Maximilien Brice (c). Dreamstime.com:
Lineartestpilot (c/scientist). 188 Science Photo
Library: Mehau Kulyk (t)

All other images © Dorling Kindersley
For further information see:
www.dkimages.com